The Metamorphosis of the Electron
Christian Holzapfel

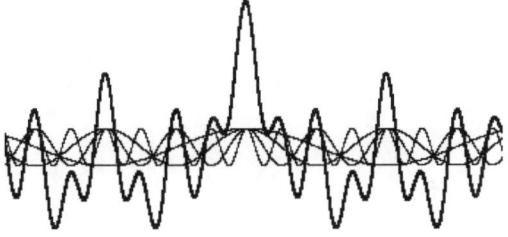

(An excursion in the landscape of physics)

for Inge

The Metamorphosis of the Electron

Introduction.	**7**
Acknowledgements.	**9**
1st day, development of the electron.	**11**
2nd day, particles, waves, and special relativity.	**29**
3rd day, classical electrodynamics.	**43**
4th day, antimatter, general relativity, and gravitation.	**69**
5th day, polarization, radioactivity, and EPR.	**95**
6th day, the identity of the electron.	**113**
7th day, polarized photons.	**117**
8th day, about laws of nature.	**129**
References.	**147**

Introduction.

The land of physics presents a bizarre landscape with many intricate pathways. Through this land we shall make a small excursion of eight days time during which we shall walk on very different paths. Some of them are broad and easy to manage. Others are dead-end streets; we shall avoid them. Some of the paths are rocky and steep; they are hard to clamber up. But the difficult climbing you will yield a rich award because the view of cognition is overwhelmingly beautiful.

The land has no end and it is not yet entirely explored. We don't even know how much there is or what new horizons are ahead. So let us start with the walk and try to explore some of this land. This book is not a textbook. It is a small book that is meant to give you a feeling for how physicists think and to show you the strange beauty of the world behind mathematical formulas that are largely not given here.

On our walk through this landscape we shall follow the life of the electron from its detection as a small, electrically charged ball to its fading away as a wave.

Acknowledgements.

Above all I want to thank my wife Inge for the correction of the German text. She, of course, knows every comma, every full stop, and all the strange words we use in physics.

Later on came the translation into English. I thank my son Rupert and our friend Gerhard Nägele from the Research Center Jülich for the work they did. The most important work was done by our friend Aaron Galonsky of Michigan State University who smoothed my clumsy English into readable English and made several useful comments. I thank him for that.

1st day, development of the electron.

When the electron was discovered it was a ball, a little ball with an electric charge *e* and with a radius r_e. The electron also had a mass m_e which was determined from the deflection of its path in a magnetic field. How the electron could exist as such a ball was unknown. Electrostatic forces should break it up immediately into an infinite number of tiny pieces. What was holding this charge together?

Then it was discovered that the electron is rotating around its axis; it is spinning. By this rotation, a magnetic field is created. Thus, the electron not only possesses charge and mass but it is also a tiny magnet. That was understandable because a rotating electrically charged ordinary ball creates a magnetic field in its surroundings. However, the electron's magnetic field shows a strange peculiarity: it is twice as big as expected from a spinning charged ball, as if the electric charge rotates twice as fast as the mass of the electron. This would mean that owing to the faster rotation of the charge, there is an electric current looping inside the electron.

From its mechanical spin and from its mass it was possible to estimate the radius of the small ball. Knowing the radius, the rotational speed was estimated to be about 2×10^{25} revolutions per second. From this value we can calculate that any point at the equator on the surface of the electron has a velocity of about 2×10^{12} cm/sec. This huge velocity led to serious problems with respect to the theory of relativity, that only allows velocities lower than the velocity of light, i.e., lower than 3×10^{10} cm/sec. The electron thus rotates almost a hundred times faster than the allowed maximal speed.

Violating the speed limit is not the only difficulty with the electron. As you will see below, the mass of the electron also makes trouble. Just as there is a magnetic field associated with magnetic poles, there is an electric field associated with electric charges. If we calculate the energy of the electric field of the electron, we can obtain its mass according to the theory of relativity. We call that mass the electromagnetic mass m_{elec} of the electron. Alternatively, if we calculate the momentum (mass times velocity) of an electron moving with its attached electric field we also get its mass m_{elec}. The trouble is that these two methods give masses that are different, and they differ by a factor of ¾; the mass calculated from the momentum is bigger than the mass calculated from the theory of relativity. So we cannot just discuss it away. We don't know which of them is the observed mass m_e mentioned at the beginning. There must be something wrong with our model of the electron.

If the state of motion of a particle with charge is changed – slowed down, accelerated, or redirected – electromagnetic radiation, such as light, is emitted (in German, *"Bremsstrahlung"*). Therefore, we need more force for the acceleration of a charge than we need for the acceleration of a neutral object with the same mass; otherwise energy would not be conserved. The rate of work against this additional force is equal to the rate of emitted electromagnetic energy. The origin of this additional force in a space which contains only an electron and an external accelerating field must lay in the electron, in an internal action of the electron on itself. The emission of electromagnetic radiation appears to come about as if there were an internal force acting on the electric charge. We call that force the *radiative reaction force*.

This additional force acts like an inertial force on a mass μ, which has to be added to electromagnetic mass. The total mass m_e that is relevant at a change of motion thus consists of at least two parts, the "field-free" mass m_{elec}, that corresponds to the mass of a neutral object which doesn't emit Bremsstrahlung and the mass μ that comes from the internal force due to the electron's charge. The field-free mass m_{elec} conflicts with the theory of relativity but is consistent with classical electrodynamics – although the theory of relativity is also consistent with classical electrodynamics. The mass μ due to internal forces depends on the internal structure of the electron, which we don't know anything about. We only know the sum of both parts. There is something very wrong with our model of the electron. A new idea is needed.

The realization came in the early part of the twentieth century that the electron bound in an atom was responsible for the emission and absorption of light. The picture developed by *Niels Bohr (1885 – 1962)* that the electron revolves like a small planet around the atomic nucleus explains this phenomenon in a simple way. When the electron falls from a distant orbit to one that is closer to the nucleus and thus more tightly bound and of lower energy, the energy difference is emitted as light. If, instead, light is absorbed from a surrounding radiation field then the electron is lifted up from an orbit of lower energy to one of higher energy. Total energy is always conserved.

Why the electron remains in the closest orbit without falling down into the nucleus was not known. Of course, it is possible to calculate the speed of the electron in the orbit, like that of the planets orbiting around the sun. This is the speed necessary to keep the electron from falling down, but, unlike the planets, the revolving electron is charged, and because its direction of motion is continuously changing it is accelerating and should, therefore, lose energy continuously by the emission of Bremsstrahlung – in the form of light – and thereby spiral down into the nucleus very quickly. Since this does not occur, it was necessary to postulate stable orbits of the electron around the nucleus, i.e., orbits from which no energy is emitted, an artificial conception of the atom which was thrust upon us by nature (Fig. 1.1).

This demand on our imagination, as formulated by Niels Bohr, was the first step to a theory with completely new ideas, the *quantum theory*. It was developed in the twenties of the last century. Quantum theory turns out to be a very fruitful instrument to predict and quantify the behavior of atoms, unfortunately without any understanding about why it works so well. For the benefit of a method of calculation by which we can predict how nature behaves we are left without an understanding of why it behaves as predicted. Quantum theory is

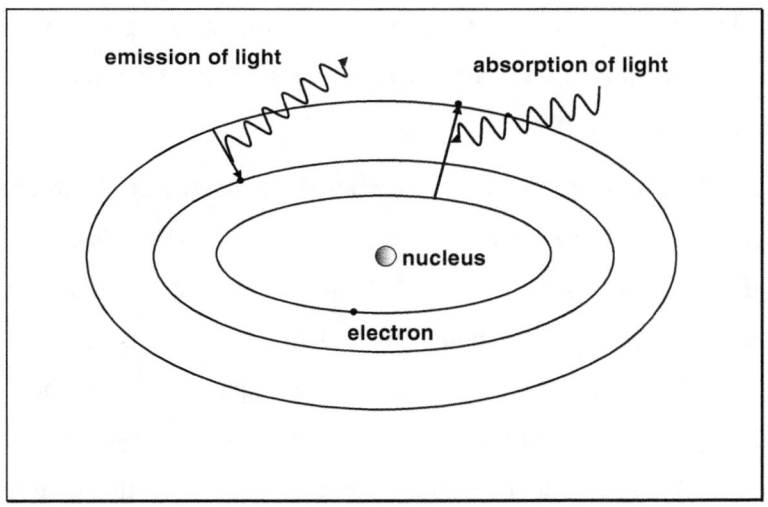

Fig. 1.1 The atomic model of Niels Bohr.

The electrons revolve around the nucleus like the planets around the sun. Three of the many possible stable orbits are shown in the figure. The electron's orbit can change from one far from the nucleus to one nearer to the nucleus, and when it does, light is emitted. When light is absorbed by the atom, the electron jumps into a more distant orbit.

represented by the famous equation of *Erwin Schrödinger (1887 – 1961)*, an equation that leads to a description of the physical and chemical behavior of everyday life, as well as of the microscopic atomic environment.

The Schrödinger equation describes the behavior of electrons in the electric field of the positively charged nucleus. With this equation it is also possible to describe the chemical behavior of atoms and molecules – in principle, although the calculations in most cases turn out to be very difficult. One important consequence is that the old periodic system of the elements is set on a sound physical basis. Unfortunately, a simple mechanical picture is lost in the quantum description because the electron appears not as a particle but as a wave, and a very odd wave, indeed.

From electrodynamics we are used to dealing with waves. But the quantum mechanical electron waves are not waves with measurable properties such as the electric and magnetic fields of light waves. Instead, the electron waves are *probability waves*. The behavior of the electron is described by a complex-valued mathematical function, the wave function, from which real numbers can be developed. Those numbers give the probabilities of the possible behaviors of the electron. The spatial probability field gives the probability to find the electron at a certain location. This is a purely mathematical description, but nevertheless, these probability waves show physical features.

Experiments with electrons passing through small slits show interference phenomena. Only waves are able to produce interference, that is, one wave can either add to or subtract from another wave. Particles are not able to produce interference, at least macroscopic particles obeying the laws of classical physics, like cannon balls, snowballs and whatever can be thrown. But electrons clearly show interference, hence, electrons are waves. They are not only described by a wave function, they are themselves waves; each electron produces one single flash on the screen. A beam of electrons falling upon two neighboring slits show an interference pattern on the screen, as if waves were travelling through both openings. One part of the wave going through one of the openings interferes behind the opening with the other part going through the second opening (Fig. 1.2).

Such interference phenomena can also be observed with normal water waves. When you throw two stones simultaneously into a quiet lake, you can see how the waves spread out in circles from the two points where the stones have plunged into the water. When the waves meet the peaks and valleys of each wave are added. Two peaks coming together create a peak double in size. A peak from one wave and a valley from the other wave compensate each other, so that the surface stays flat. And where two valleys meet a deeper valley results. The whole story moves with the wave fronts.

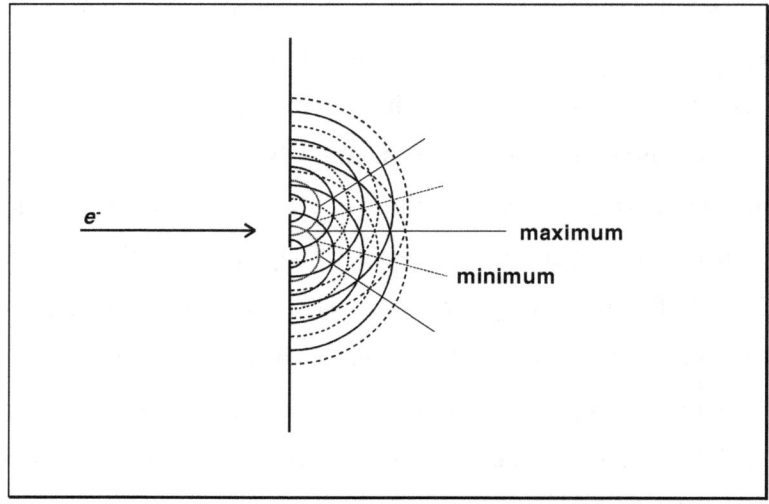

Fig. 1.2. Two-slit scattering experiment.

By the diffraction (the apparent bending of waves around small obstacles and the spreading out of waves past small openings) of electrons at a double slit, interference patterns are created akin to water waves or light waves. Where the wave peaks (solid circles, wave fronts) meet we find the maxima (solid straight lines); where wave peaks of one field come together with the valleys of the other field (dotted circles) we have the minima (dotted straight lines). (Where the valleys of the two fields come together we also have maxima).

Thus, a nice interference pattern is created on the surface of the lake. This is typical for waves. Exactly such an interference pattern is seen when one electron is allowed to go through either of the two openings. The openings must, however, be very close together because the wavelength of an electron is very short.

In fact, in 1961 the experiment was carried out by a young physicist, *Claus Jonsson (1930 –)*, a student at that time. He used thin metallic foils with slits – smaller than 1 μm, size and separation.

In 1801, with a similar two-slit experiment *Thomas Young (1773 – 1829),* showed that light is a wave, as already proposed much earlier by *Christian Huygens (1619 – 1695)*, an electromagnetic wave. Also for light, the two openings should be very close in order to observe an interference pattern.

The interference pattern from the electrons coming through the two slits shows up as bright stripes on the screen. These stripes are the result of all the single flashes created by individual electrons. The sum of all these flashes creates the interference pattern. But each single flash shows that the electrons are hitting the screen like bullets. Thus the electron is clearly both particle and wave, a particle which can be localized at a certain point in space (on the screen) and which has a certain velocity, and a wave filling the whole space – crazy but true.

If you produce a wave system consisting of waves of several wavelengths then all these waves are superposed. At some point in space they momentarily amplify each other when wave hills coincide and then cancel each other out when wave hills and valleys meet. This phenomenon is well known to radio amateurs as so-called beat. When they try to receive a radio station using a frequency near to that of another station, the reception may be good for a certain time; but then the station signals fade away exactly at the moment when the radio speaker is going to tell something of interest; a few seconds later the speaker appears again. One can hear the beat with a frequency very much lower than the frequencies of the two stations disturbing each other (the beat frequency is the difference of the two station frequencies).

That's how the so-called *wave packets* arise. The electron can be interpreted as such a wave packet. However, we should keep in mind that these waves are probability waves, and that means the wave packet is in the place where the probability of finding the electron is high, (Fig. 1.3.a).

Maybe one should be careful and say that the electron is not a wave, it just behaves like a wave – but at the same time it is not a particle, it just behaves like a particle, depending on how we have designed our experiment.

Such a wave packet has a peculiar feature. To each wave in the packet we assign a velocity equal to the velocity of the particle described by that single wave. We just say the wave has a certain velocity. According to *Louis de Broglie (1892 – 1987)* the velocity of the probability wave is connected to its wave length – unlike the normal waves of sound or of light.

The wave packet consisting of several waves is also moving with a certain velocity, the so-called group velocity. It turns out that this group velocity is equal to the velocity of the particle associated with the packet. But because there are several waves and thus several particle velocities involved, the group velocity is not exactly defined; that means also the velocity of the particle associated with the wave packet is not exactly defined. At the same time the wave packet also has a certain extent Δx *(delta x)*, so the location x of the wave packet is also not exactly defined.

The more exactly its location is determined, i.e., the narrower the wave packet, the more waves of different velocities are required to compose it, i.e., the greater is the uncertainty in its velocity, (Figs. 1.3.a and b).

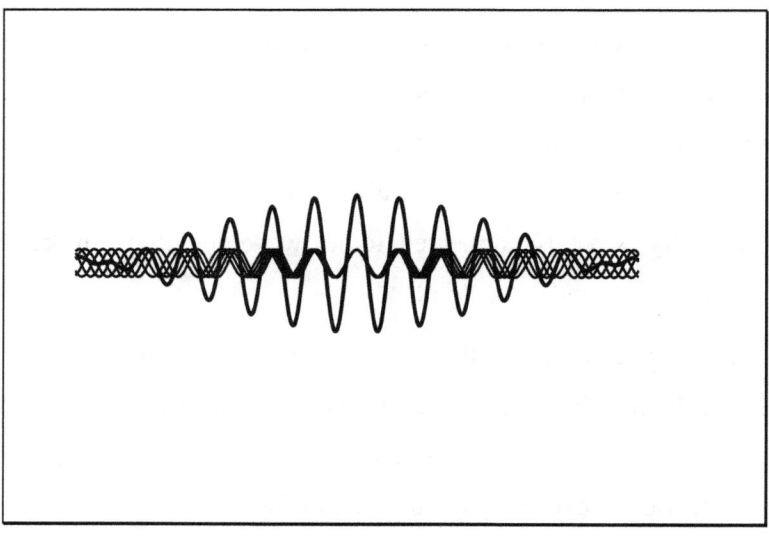

Fig. 1.3.a. Wave packet in space at a certain time.

The superposition of waves of different wavelengths (thin lines) produces a wave packet (thick line). In this figure the wave packet consists of five waves with different wavelengths. The wavelengths of these five waves correspond to phase velocities ranging from 5.6 to 6.4 arbitrary units (a.u.). That means the uncertainty of the group velocity Δv of the wave packet is 0.8 a.u. Compare this to the wave packet in Fig. 1.3.b, where the wave packet consists of nine waves with different wavelengths.

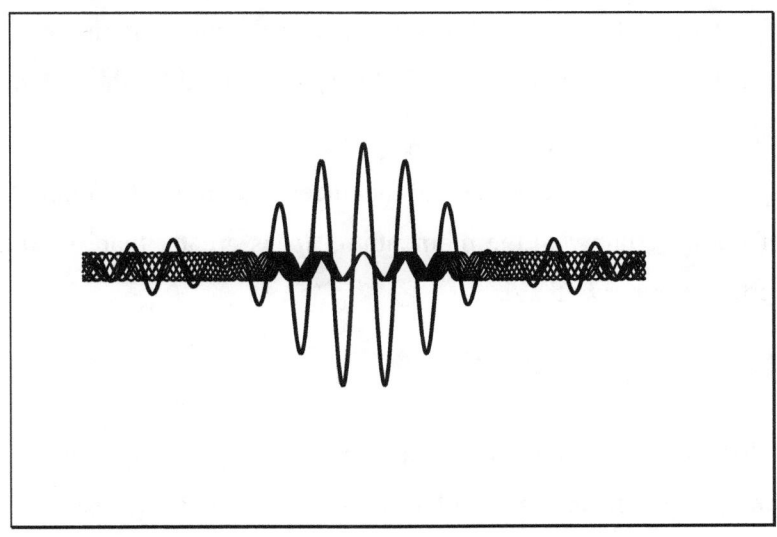

Fig. 1.3.b. Wave packet with more waves.

The superposition of nine waves of different wavelengths produces a narrower wave packet (thick line). The wavelengths of these nine waves correspond to phase velocities ranging from 5.2 to 6.8 arbitrary units (a.u.). That means the uncertainty of the group velocity Δv of the wave packet is 1.6 a.u., i.e., two times the uncertainty of the velocity Δv in Fig. 1.3.a. This wave packet, therefore, is much more concentrated around the center, i.e., the uncertainty of the location Δx is much smaller, about half the uncertainty of the location Δx in Fig. 1.3.a.

The product $\Delta v \times \Delta x$ has almost the same value in both figures. (Δx is taken as the half-width of the packet, i.e., the width at half maximum).

In place of the velocity, we can consider the momentum p of the wave packet, which is simply the velocity multiplied by the mass of an electron.

It turns out that the product of the uncertainty in the location, Δx, and of the uncertainty in the momentum, Δp, is greater than or equal to a certain number h, that is

$$\Delta x \, \Delta p \geq h$$

If the location of the electron is known exactly, $\Delta x = 0$, then Δp is infinitely large, hence, the velocity is totally undetermined: The perfectly localized wave packet consists of waves of all possible velocities from zero to infinity. On the other hand, if the velocity is exactly determined, $\Delta p = 0$, then the wave packet consists of one single wave of exactly known velocity. The wave is uniformly spread out in the entire space without forming a packet; in short, its location is completely undetermined so that Δx is infinitely large.

This is the famous *uncertainty relation* found by *Werner Heisenberg (1901 – 1976)*. The quantity h is a universal constant known as *Planck's number* (in German: *"Wirkungsquantum"*). The uncertainty relation is valid for our wave packet, which is a mathematical construction. The electron behaves exactly in the same way as the wave packet – only we don't know why.

In summary, the equation of Schrödinger describes the position of an electron in the form of a probability wave. The uncertainty relation is formulated for waves of probability and also describes the behavior of the electron. If we try to confine the electron in a small volume, e.g., in the interior of an atom, then its velocity will be uncertain by an amount described by the relation of Heisenberg, i.e., the classical picture of the orbital movement of an electron (a tiny ball) with a certain velocity around the nucleus has no physical meaning. The electron rather seems to have a range of velocities and to be smeared out in the neighbourhood of the nucleus (Fig. 1.4).

In fact, the Planck number h is so small that the Heisenberg relation plays no role in our everyday life. We always know where we have our stuff – and if the location of our things is uncertain, then it has nothing to do with Heisenberg.

The uncertainty relation also prevents the electron from falling into the nucleus, because if the electron would be located in the tiny space of the nucleus it would have such high velocity components that it would be catapulted out again. Due to the uncertainty relation, the closer the electron comes to the nucleus the higher is its kinetic energy and the more tightly it is bound to the nucleus. Another way of expressing the tighter binding is to say that its potential energy in the electric field of the positively charged nucleus is more negative. The system electron + nucleus will adjust to a state where the total energy, the sum of the two kinds of energy, kinetic and potential, acquires its lowest value.

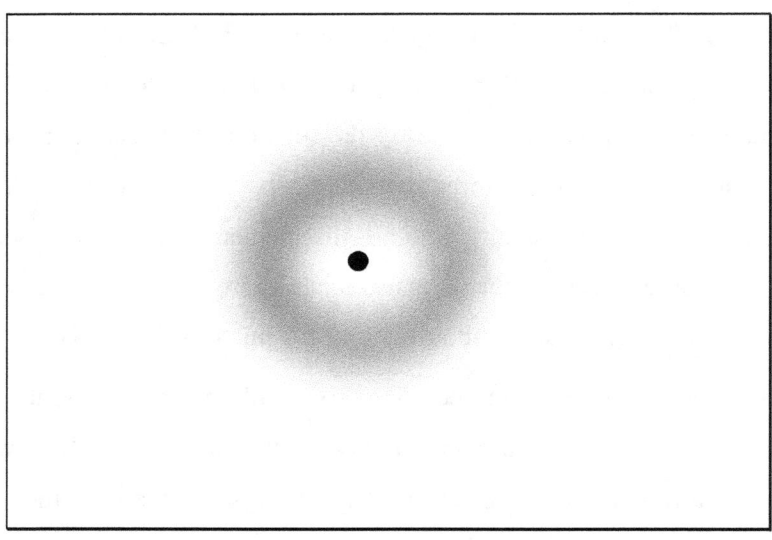

Fig. 1.4. The smeared out electron.

The electron is not a particle located at a fixed point on a well-defined orbit but is smeared out in the whole neighbourhood of the nucleus.

The cloud represents the probability of finding the electron at various places around the nucleus.

You may now get the idea that there is a way to trick the uncertainty relation. If you follow the wave packet of a freely moving electron for a long time you can measure the velocity as exactly as you want just by dividing the distance travelled by the time; the longer you wait the more precisely is the velocity known, the smaller is Δv. Thus, the product $\Delta x\ \Delta v$ also becomes smaller and smaller. But unfortunately the wave packet is not stable. Because it consists of waves with many wavelengths it is subject to what is called dispersion. The waves are running apart, the packet disintegrates. (Figures 1.3a and b were for a moment in time.) For a free electron with a certain velocity spread the wave packet may have a width of about 10^{-10} m according to the uncertainty relation. After one second the width is more than 1000 km. So there is no possibility to make the product smaller; nature cannot be deceived.

By the way, it is a general principle that a physical system tries to reach a state of lowest energy of some kind; that's why houses of cards collapse, pencils fall down, and electrons, according to the classical concept, fall into the nucleus. In all these cases it is the potential energy that is minimized. According to the uncertainty relation the electron will be found with the highest probability where the total energy is a minimum. That is the most stable state of the system nucleus + electron, the atom. The space occupied by the electron, that is, the size of the atom, will adjust so that the total energy is a minimum. Therefore, the size of the atom is a direct consequence of the uncertainty relation.

In classical electrodynamics the energy that strives for a minimum is only given by the potential energy of the electron. Therefore, the classical electron falls into the nucleus. By the uncertainty relation or by the description with the equation of Schrödinger the kinetic energy is added, and that prevents the collapse of the atom.

2nd day, particles, waves, and special relativity.

In 1924 Louis de Broglie postulated that matter consists of waves, so called *matter waves*. Not only electrons, but all bodies: protons, neutrons, atoms, and also snowballs, even the celestial bodies, earth and moon. They all consist of waves. By one simple equation ($p = h/\lambda$) he related the momentum of the body to the wavelength of these waves. This idea of de Broglie was first rejected by the Sorbonne University in Paris as nonsense.

A few years' later diffraction phenomena were demonstrated with electrons moving at high speed, and a little later even with whole atoms like helium and with molecules like hydrogen. Such diffraction phenomena are also interference phenomena, as they occur at edges or at small holes or slits or in crystals, phenomena that are typical for waves. Hence, it was shown experimentally that matter in general, not only electrons, behaves like waves. Therewith, it was also nicely possible to explain why certain orbits of the electron moving around the atomic nucleus were stable, which means did not emit radiation. These stable orbits are just those where an integer multiple of the wavelength fits in the orbit's circumference (Fig. 2.1).

Stable orbits without radiation means that the electron on such an orbit does not lose energy by radiation and does not fall down into the nucleus. The system of electrons around the nucleus, that is the atom, can still emit radiation by jumping of electrons from one stable orbit to another (lower-energy) stable orbit.

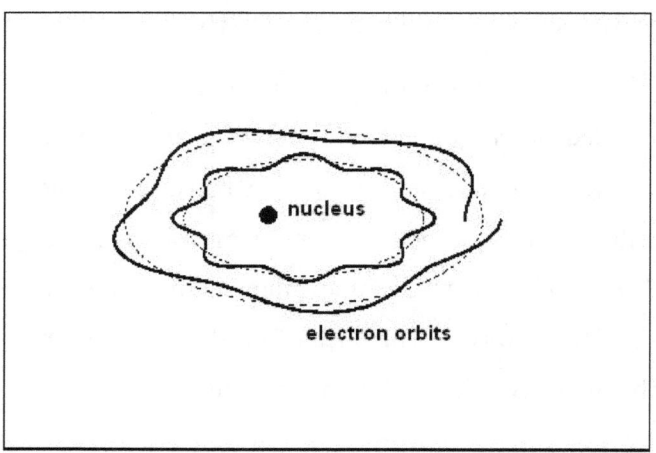

Fig.2.1. Stable and unstable electron orbits.

The orbit closer to the nucleus is stable; here an integer multiple of the wavelength fits in the orbit, namely 8 waves in this example.

In contrast, the outer orbit is not stable; here the waves do not fit in the orbit.

These stable orbits are the only allowed orbits for the electrons. So it looked like the primary idea of Niels Bohr had been proven by means of the Schrödinger equation and the uncertainty principle. But that was not the case; rather, the stable orbits with the integer number of waves was only a smart formulation. A consistent formulation had been developed based on the idea of Niels Bohr, so that this idea about the stable orbits followed causally – otherwise the formulation wouldn't be consistent. The question, why it was consistent, was still unanswered. The pictures fitted somehow together, only why they fitted so nicely, why for example the orbits were radiationless, was still without answer. And what matter waves consisted of, what was oscillating, were questions that remained totally unclear – again our mind was overstrained. The picture will be revised later on our trip; it is still a semi-classical picture.

The bigger the bodies are, the more the wave properties disappear. In our everyday life we meet macroscopic bodies: billiard balls, soccer balls, and cars. They don't show wave properties at all. Only small bodies such as electrons and atoms show wave properties.

A little earlier, in 1900, *Max Planck (1858 – 1947)* developed the idea that light, supposed to be a wave, also consists of particles, called *photons*. With the same relations that were used by de Broglie to combine waves with particles, electromagnetic waves were combined with photons, also called *light quanta* – a quantum is just something very small, although in politics sometimes the term quantum leap is used to describe a big change.

Albert Einstein (1879 – 1955) took up an old idea of *Sir Isaac Newton (1643 – 1727)* about light. Newton had the idea that light was a stream of particles. That idea, as we mentioned yesterday, later seemed to be disproved. The interference phenomena shown by light indicated that light is a wave. But new experiments, where electrons could be kicked out of metallic material, showed that light also has particle properties.

We recognize two anti-parallel developments of ideas. Electrons are turned into probability waves and light waves are turned into photons. Of course, there are some differences, but both of these developments demonstrate that the phenomena we got to know cannot be explained with just one of the conceptions. To explain the behavior of electrons we need both ideas, the particle picture and the wave picture, and to explain the behavior of light we also need both ideas. These two pictures cannot be combined into one consistent complete picture; both of them are necessary. This fact, that we need both pictures in order to describe nature adequately, was formulated by Niels Bohr as the *complementary principle*. The two pictures are complementary.

The present view is that particles, quanta, can act like waves, and waves, electromagnetic waves, can act like particles, photons. Both kinds of particles, electrons and photons, can appear as waves in diffraction phenomena, and both kinds can appear as particles. The matter wave is interpreted as a probability wave.

In 1913 *James Franck (1882 – 1964)* and *Gustav Hertz (1887 – 1975)* demonstrated that photons, like billiard balls, could be pushed by electrons, and vice versa, in 1922 *Arthur Compton (1892 – 1962)* demonstrated that electrons could be kicked by photons. Electrons and photons both behave like tiny billiard balls.

But these contradictory pictures both have to correspond with reality, above all with nature, with classical physics as we experience it macroscopically. Indeed, the bigger the system described, the more classical features the quantum theory exhibits, the uncertainty relation becomes unimportant, and interference phenomena and other wave properties disappear. Nobody has ever seen interference phenomena with tennis balls. Ultimately, the results of both conceptions, classical physics and quantum theory, become identical, although the concepts are basically different. We shall come back to this with an interesting example later, on the 4th day.

The merging of the two contradictory pictures was formulated by Niels Bohr as the so-called *principle of correspondence*. Quantum events correspond to classical events in a way that the results of quantum theory converge towards the corresponding classical results when the quantum number increases (Fig. 2.2).

Now we have met a new term on our walking trip, the *quantum number*. That is nothing more than a numbering of the possible states in which a system can exist, for example, the possible electron orbits in an atom. The orbit closest to the nucleus, the orbit with the lowest energy, gets the number 1, the quantum number 1. The next larger orbit gets the quantum number 2, and so on. If an electron is jumping between such orbits, light is either emitted or absorbed, but only with the energy difference between these two orbits. There is nothing between the orbits. The energy is emitted or absorbed as quanta, as photons, small packets with this well-defined energy. In the classical theory of electrodynamics there is no movement of the electron by which it is possible to calculate these energy packets. Any attempt would lead to senseless results. It is not the jumping electron that emits light; the whole system emits light when the electron is jumping. It has to be described by the quantum theory.

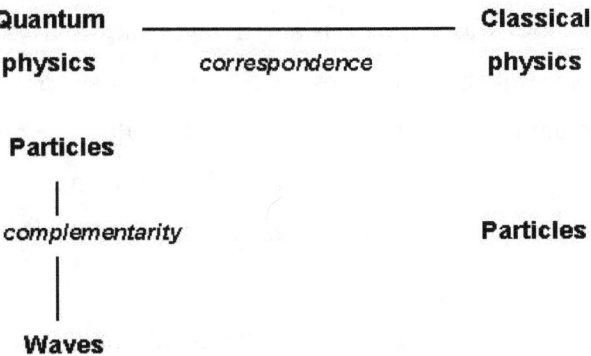

Fig.2.2. The two principles formulated by Niels Bohr.

Quantum theory merges into classical physics for large quantum numbers. Phenomena and incidents in classical physics correspond to phenomena and incidents in quantum physics.

Particles and waves are complementary. That means they complement each other in describing the behavior and properties of the electron. Both pictures are necessary for a complete description.

If we go to very high quantum numbers, that is to electrons moving in orbits far away from the nucleus, then it is possible to treat the electron motion as if in a small antenna, and classical electrodynamics yields the correct results for emission and absorption. Of course, quantum theory also gives the correct results. The results of classical electrodynamics and of quantum theory do not differ, or more exactly, the differences become smaller and smaller the higher the quantum number is.

This picture now looks very nice. For small quantum numbers and for small systems quantum theory is appropriate, and for high quantum numbers and big systems classical physics is appropriate. The principle of correspondence makes sure that there is no break between the two theories. The change from one theory to the other is smooth and steady.

But the nicer the picture becomes the more questions appear, and looking at the picture more closely many things become incomprehensible.

When an electromagnetic wave propagating in space in all directions as a photon hits a single electron, the total energy of the wave field suddenly has to concentrate on the electron. How that is managed by nature remains a mystery. The sophisticated term *the collapse of the wave function* is used, but the meaning is only "*???*".

The same question turns up with the electron interference experiment of Fig. 1.2 when each single electron triggers a flash at the point where it hits the screen, and only after many of these flashes do we see the interference pattern and have evidence that there is a wave field. Again, the probability wave has to concentrate at one single point every time an electron hits the screen.

A severe flaw in the quantum theory we have so far described is that the Schrödinger equation does not agree with the theory of relativity.

In the 19th century it was already known that light propagates like a wave, an electromagnetic wave in which electric and magnetic fields oscillate. But the medium, in which these waves propagate, the so-called ether, was strange and unknown. The waves' velocity of the propagation was first measured by *Olaf Römer (1644 – 1710)* in 1676. He came amazingly close to the now known value of 300,000 km/sec with his observation of the moons of Jupiter. That enormous speed means seven times around the earth in one second.

This medium, the ether, had to have odd properties. In order to transmit light with such a high velocity it should be harder than steel. On the other hand, the celestial bodies – planets, sun, even the galaxies – should be able to move unhindered through this ether.

At the end of the 19th century *Albert A. Michelson (1852 – 1931)* and *Edward Morley (1838 – 1923)* made their famous interference experiment in order to determine the velocity of the earth through this ether, a velocity meant to be the absolute velocity in space. Our globe travels with 30 km/sec as it orbits the sun. That speed would take us from Jülich to Aachen in one second, or the length of Manhattan in

0.7 seconds. At the same time our sun, along with the whole system of planets, is travelling through the ether in our galaxy, and the galaxy rotates in this ether. Despite all these motions it is understandable that scientists wanted to state the absolute velocity through space. At least it should be possible to observe the difference of 2x30 km/sec between light moving in the same direction as the earth and in the opposite direction.

They found no difference! And from this negative result Einstein developed his theory of relativity.

Einstein generalized from the experiment that the velocity of light was the same in all inertial systems, i.e., in all systems moving with a constant velocity. That means light's velocity has one value (in a vacuum) and is independent of the speed of the system in which it is measured. He further concluded that there is no such thing as an absolute velocity, all velocities are relative. It is only possible to measure a velocity of a body relative to some other body.

As a consequence, some other things also proved to be different from our everyday life beliefs. The idea of an absolute elapsing time turned out to be wrong. Every system has its own time. Also, the size of an object became dependent on the velocity of the observer. All became relative. The Michelson-Morley experiment is an example of how a negative result, the "failure" of an experiment, can lead to a totally new view of nature.

Instead of the old familiar *Galilean transformation* (the conversion of coordinates and other parameters between two systems moving with different velocities) we have to use the so-called *Lorentz transformation*.

In the Galilean transformation velocities are just added. That means if I walk at 5 km/h towards the restaurant car in a train that is moving at 80 km/h, then I move at 85 km/h relative to the landscape outside.

To add these two velocities with the Lorentz transformation we also have to consider the ratio between the velocity of the train and the velocity of light. Then my velocity relative to the landscape is a little less than 85 km/h. There are more such strange effects, but the differences are so small that we don't feel them in our everyday lives. Only if the velocity of the train could come close to the velocity of light would these relativistic effects be easily noticed.

But they are always present. And in a theory about nature, also about the nature of the electron, we have to take them into account, particularly when the velocities we are dealing with are very high.

When we add velocities with the Lorentz transformation it turns out that the velocity of light is also the maximum possible velocity. If we are sitting in a train moving with almost light velocity – what fun! – say 295,000 km/sec, and I turn on my flash light in the direction of motion and send a light beam with 300,000 km/sec ahead, then this light beam only moves with 300,000 km/sec relative to the landscape and not with 595,000 km/sec.

The Lorentz transformation produces Einstein's conclusions that in all inertial systems the speed of light is the maximum possible velocity and it is a constant of nature.

Several ideas we got used to over the millennia we had to abandon. Events, which are simultaneous in one system, are not simultaneous in another system moving relative to the first. Clocks are running differently in different systems. But that is another story. Anyhow, maybe we should have a look at the story with the clocks because it demonstrates how curiously nature is treating our feeling of logic.

If I am standing with my pocket watch at the roadside observing a car passing, I notice that the clock in the car is running more slowly than my watch. That may be understandable. But the theory of relativity states – as the name says – that all is relative. The car driver is right when he argues that his car doesn't move but instead the whole earth, including my watch, is moving backwards under his wheels, and he therefore notices that my watch is running more slowly than his clock. He is right and I am right. The contradiction comes from our unconscious imagination of a universal clock with which we compare our watches. This universal time just doesn't exist. Each system has its own time and we cannot compare them absolutely. That's how nature is, even if we feel it is not logical.

However, we can only observe these effects if the car is very fast, near the speed of light. In normal traffic on the road we can trust in the Galilean transformation. But with devices that accelerate elementary particles we achieve velocities close enough to the speed of light so that the slower clock of those particles, their time dilation, is observable.

Before we come back to the electron we should have a glance at one important implication of relativity theory that we shall face later on in connection with the electron.

Until the beginning of the 20th century energy and mass were considered to be two different things. However, with relativity theory Einstein showed that these were manifestations of one thing. Energy *E* and mass *m* can be converted into each other by the simple formula

$$E = m\ c^2$$

At first sight it doesn't say very much, it is just a multiplication with the square of the light velocity *c*. In 1945 this formula became a terrible reality in Hiroshima and Nagasaki.

The formula tells that we can turn mass, for example a lump of sugar into energy. That energy is not only the chemical energy of sugar, which is about 10 Calories, about the energy you work off in one minute on a bike. It is the energy of the entire mass of the lump. We would get 50 billion Cal according to the formula. That is about what all homes in the city of Jülich consume in one year in the form of electricity – all from one lump of sugar.

That is the energy converted from mass in the nuclear power plants. There, of course, not sugar is used but uranium – anyway mass is mass. In fact, only a tiny fraction of the uranium is converted, but from 1 kg of uranium fuel we gain as much energy as we gain by burning 85 metric tons of hard coal – about three wagons filled with coal.

But now back to the electron and to its relative, as well, to the photon. They have a particularly close relationship, as we shall see.

3rd day, classical electrodynamics.

In 1864 *James Clerk Maxwell (1831 – 1879)* wrote down the equations that describe the behavior of light – only four short equations (Fig. 3.1). They describe the behavior of all electromagnetic waves. It is a system of equations, which, with their beauty and simplicity, show an ingenuity that makes one think a god had written them down, like the Ten Commandments Moses was given on Mount Sinai.

For the layman they look like hieroglyphs, but for one who can interpret them they open a whole world. Maxwell's ingenuity was seen to its full extent when half a century later it turned out that this system of equations was Lorentz invariant, i.e., the equations do not change under a Lorentz transformation. Hence, classical electrodynamics is consistent with the theory of relativity.

These four equations, together with some few addenda, describe the whole of electrodynamics, that is, the theory of magnetism and of moving electrical charges. When we deal with the electron it is, therefore, appropriate to have a look at electrodynamics. It is, so to say, the playground of the electron. The electron and its two relatives, the positron and the photon, are important actors in this play. Later on, on day 4, we shall deal with the positron.

$$curl H = \frac{1}{c}\frac{\partial}{\partial t}E + \frac{4\pi}{c}j$$

$$curl E = -\frac{1}{c}\frac{\partial}{\partial t}H$$

$$div H = 0$$

$$div E = 4\pi\rho$$

Fig. 3.1. Maxwell's equations.

H is the magnetic field, **E** is the electric field, **j** is the current density, **ρ** (rho) is the density of charge, and **c** is the velocity of light.

The first equation describes the magnetic field induced by a changing electric field and by an electric current.

The second equation describes the electric field induced by a changing magnetic field.

The third equation says that the magnetic field has no sources, i.e., there exist no magnetic monopoles, the magnetic field lines are closed curves.

The fourth equation says that electric charge is the source of the electric field.

You may remember electrodynamics, just called electricity, from physics lessons in school as a confusing mess of phenomena with electric wires, coils, magnets, and much more; and terms like fields, induction, Lorentz force, and Hall effect. The task of a physicist is to work out the basic principles so that it is possible to see the scheme in which all these phenomena can be arranged – like Isaac Newton did when he understood that the apple falling from the tree and the movement of the moon both were caused by the force of gravity.

In order not to make the description of the phenomena too complicated, it is useful to have a look at some of the terms used. First of all, it is important to state that electrodynamics is a *classical* theory and that all the questions and difficulties concerning the structure of the electron do not play any role. In brief, electrodynamics of the electron is the theory of its movement without regard to what an electron is. We only need to know that the electron possesses a mass and a charge – the charge of the electron being negative and that of the positron positive.

Everyone is familiar with the term *current*. A stream, or current, of water in a river or a creek is just the coordinated movement of water particles. When charges, electrons or positrons, for example, are moving together, we have an electric current. The water stream is caused by gravity – water flows from the mountain down into the valley because the earth attracts the water. Exactly like that a current of electrons is caused by the pull of positive charges. There is attraction between opposite charges (positive and negative charges) and repulsion between similar charges (positive and positive or negative and negative charges). Two electrons repel each other as do two positrons, whereas electrons and positrons attract each other (Fig. 3.2). Why they do so, nobody knows, but they do.

Of course, what is valid for positrons is also valid for protons. Protons have the same charge as positrons, only they are much heavier. Remember, protons are located in the nucleus of the atom; that's why the nucleus has a positive charge. Atoms are electrically neutral because the positive protons balance the negative electrons.

The way an ordinary battery in a flash-light is constructed is that one end, the negative terminal, has a surplus of electrons, whereas the other end, the positive terminal has too few electrons, that is, fewer electrons than protons. This surplus of protons makes that terminal positive. Protons are fixed in the nuclei of the atoms in the material, whereas electrons, because they are much lighter, are moving more-or-less freely in the material. If the two terminals of the battery are connected with a copper wire, the electrons in the wire are repelled from the negative terminal and attracted by the positive terminal; the electrons move in the copper wire, i.e., a current flows through the copper wire.

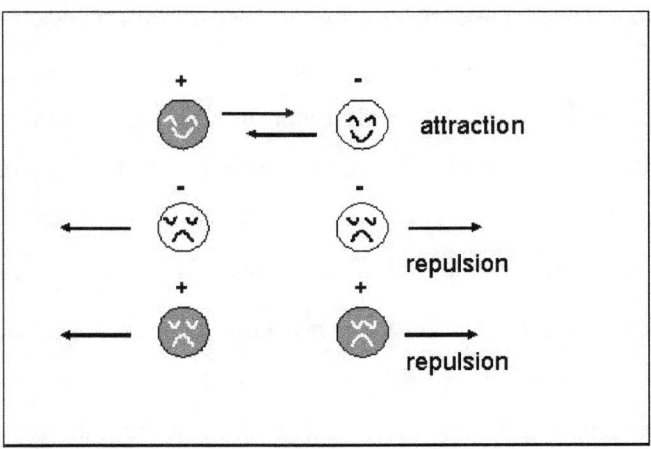

Fig. 3.2. Attraction and repulsion.
Opposite charges attract each other, like charges repel each other.

By the way, in the early days of electricity current was defined as the movement of positive charges flowing from the positive terminal to the negative terminal. Later it was recognized that the moving parts were actually the negative electrons flowing from the negative terminal, where there are too many of them, to the positive terminal, where there is a deficit of electrons.

Nevertheless, the definition was kept because looking from outside it doesn't make any difference. For almost all phenomena it doesn't make any difference if positive charges are moving in one direction or negative charges are moving in the opposite direction.

In 1820 the Danish physicist *Hans Christian Ørsted (1777 – 1851)*, with the aid of a compass, discovered that in the surroundings of an electric current there exists a *magnetic field*. Near the copper wire in which the current flowed a compass needle was deflected. As you know, a compass needle is a small magnet with a north pole at one end, pointing towards Canada, and a south pole at the other end, pointing towards Antarctica. The needle was turned crosswise, not radially, perpendicular to the wire when the current was switched on; and when the current was switched off the needle turned back so that it pointed north again. Using that compass needle Ørsted was able to discover the spatial properties of the magnetic field.

By reflection – or to say it in a scientific way: by symmetry reflection – it is possible to tell how the magnetic field should look.
If, for example, the wire is mounted horizontally, the principle of symmetry reflection tells us that the magnetic field must be the same

above and below the wire and also left and right of the wire because the current doesn't know about up and down, left and right.

Also, it is reasonable to suppose that the magnetic field is proportional to the magnitude of the current. High current means strong magnetic field, low current means weak magnetic field – no current means no magnetic field.

In addition, it is logical to suppose that the magnetic field is weaker far away from the wire than close to the wire. All these points had, of course, to be proved experimentally.

Since the compass needle is aligned crosswise perpendicular to the wire, it is clear that the magnetic field lines are rings around the wire (Fig. 3.3) – and that is exactly what the first and the third equations of Maxwell tell us.

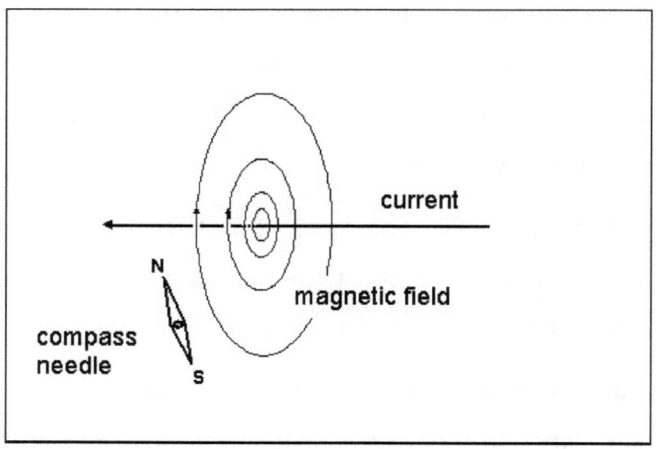

Fig. 3.3. The magnetic field of an electric current.

The magnetic field is arranged in circles around the wire where the current flows. The direction of the magnetic field is like the direction of rotation of a screw that is progressing in the direction of the current.

The third equation also tells us that no isolated single magnetic pole, no monopole, either north or south, exists. The magnetic field lines are closed curves, circles around the current.

In an ordinary bar magnet the field lines are going from one pole through the surrounding space to the other pole and back through the inner of the bar magnet to the original pole. Here, too, the field lines are closed curves. There are two possible directions of these closed lines and of the compass needle depending on which of the ways the current flows.

I am not going to discuss the mathematical details, I shall only tell you what the equations describe. If you are familiar with vector analysis you will recognize it anyway. H is the magnetic field and j is the current density. So the second term on the right side of the first equation tells us that the magnetic field lines are circles around j. The first term right we will consider later.

I have been speaking a lot about the *field*, so it is time to explain what a field is. We have already mentioned the electric field and the field of gravity; now we have heard about the magnetic field. That may sound very mysterious, but it is not more mysterious than a field of wheat. We speak about a wheat field whenever we see a large area where a wheat stalk is standing at every point. All these wheat stalks together form the wheat field where the wind in summertime plays so beautifully that you can see the waves moving across the field. Where the stalks are tall we say the field is high, and where the stalks are low the field is low.

In the same way we speak about a magnetic field if we have a region of space where a compass needle would feel a force at every point. At each point of such a field we can tell the magnitude of the magnetic force by the force with which the compass needle is deflected. We can also tell the direction of the magnetic force at each point; it is the direction in which the compass needle points. Therefore, we can draw lines – draw threads in space – so-called *field lines*, which show the direction of the force at each point. Such a field where we have both the magnitude and the direction is called a *vector field*.

These field lines have no material reality, they simply serve as a display of the vector field. I emphasize this because sometimes they are discussed as if they were real rubber straps. If you strew iron filings on a sheet of paper in a magnetic field you can "see" the field lines. In fact, what you do see is how the filings form chains or threads in the direction of the magnetic field, not really field lines.

Thereby, we should remember that we don't know what a magnetic force "really" is; we only know the magnetic force from its action on the compass needle. In the same way we don't know what the electric force or gravity "really" is. We just know these forces by their actions on charges or on masses, respectively. We can only describe these forces as given attributes of space.

Starting from Ørsted's phenomenon that moving charge (a current) creates a circular magnetic field around its path, we are able to explain other phenomena, e.g., that a coil of copper wire through which a current flows, creates a magnetic field that resembles the magnetic field of a bar magnet. With such a coil it is possible to build electromagnets. All these phenomena are based on the phenomenon discovered by Hans Christian Ørsted.

Now we also understand how a magnetic field arises from an electron revolving around a nucleus – at least in the classical picture. The revolving electron is like a one-turn coil of current around the nucleus creating the magnetic field.

Likewise, the spinning electron, which is rotating charge, creates a magnetic field. But now we are in midst of all the difficulties we saw on the 1st day. The small spinning, charged ball creates a magnetic field which looks like the magnetic field of a short coil or of a short bar magnet (Fig. 3.4).

The phenomenon that an electric current is accompanied by a magnetic field is called *Ampère's law* in honor of the French physicist and mathematician *André Marie Ampère (1775 – 1836)*, although it was Ørsted who discovered it. Ampère's special contribution is that he calculated the field.

We shall now look at a phenomenon that arises whenever an electron – or any charge, positive or negative – moves in a magnetic field.

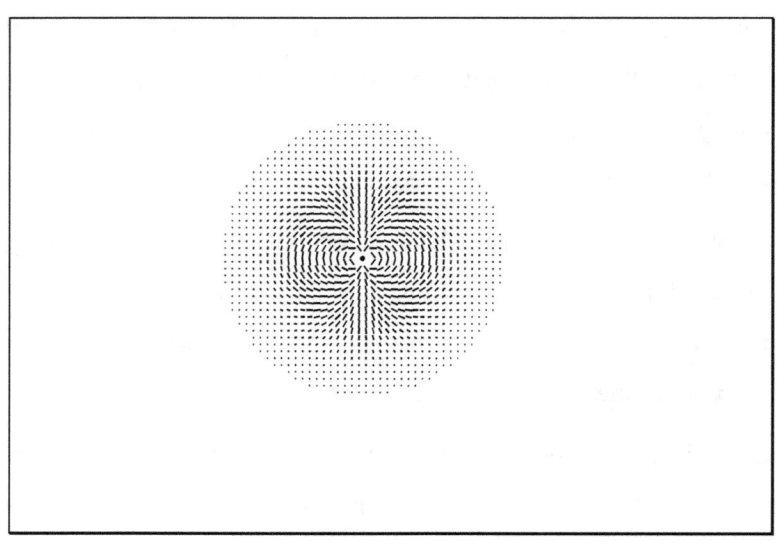

Fig. 3.4. The magnetic field of an electron.

The magnetic field of an electron is the same as the magnetic field of a small and very short bar magnet or like the magnetic field of a small flat coil with current. It is the magnetic field of an electric charge spinning about an axis lying in the plane of this page.

In this case a force acts on the electron, deflecting it perpendicularly to both its direction of movement and to the direction of the magnetic field (Fig. 3.5).

This force is called the *Lorentz force* in honour of the Dutch physicist *Hendrik Antoon Lorentz (1853 – 1928)*.

This phenomenon has nothing to do with the previous phenomenon, the Ørsted phenomenon, at least in the classical picture. To say it clearly, it is another basic phenomenon, and it is true for any charge. In classical electrodynamics this phenomenon cannot be deduced from the Ørsted phenomenon. Other physicists claim that the Lorentz force can be deduced from the Ørsted phenomenon, and it can, but only with help from the relativity theory of Einstein.

Mostly it is a matter of opinion as to which phenomenon is basic and which can be derived. Either way, the phenomena have to fit together like a jigsaw puzzle. It takes all the phenomena together to form the picture. However, with the choice of basic phenomena made here, it is easier to understand the picture. It is then also easier to search after the extension, the generalization, the higher outlook.

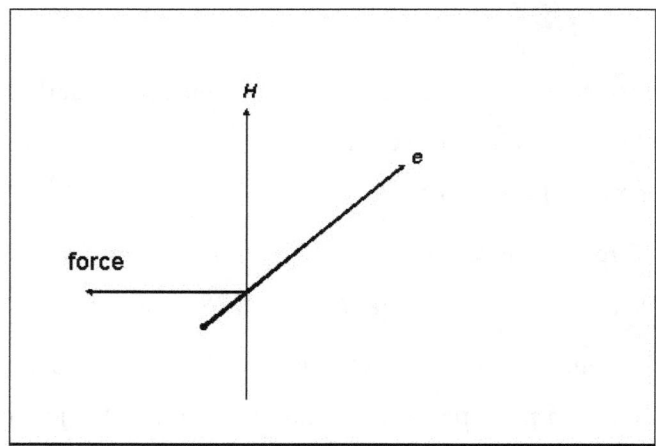

Fig. 3.5. The Lorentz force.

A moving charge – here an electron *e* – in a magnetic field **H** experiences a force deflecting it in the direction perpendicular to the direction of movement and perpendicular to the direction of the magnetic field.

You may have seen the demonstration at a physics lecture where two parallel wires in which currents flow attract each other if the two currents flow parallel and repel each other if the two currents flow anti parallel. This can easily be explained by the combination of these two phenomena, the Ørsted phenomenon and the Lorentz phenomenon. Each of the wires experiences the Lorentz force in the magnetic field of the other wire.

The Lorentz force enables us to build electric motors with which we run our vacuum cleaners and operate our electric locomotives. With the Lorentz force we also direct the electrons in our TV (at least in the old TV with cathode ray tube, CRT) to display the picture on the screen so we can follow the news and look at the crime films. Also, the display of e-mails on the screen of our PC monitor, as in the TV, is directed by the Lorentz force – although the story is different in the new LCD and plasma flat screens.

The third phenomenon we shall deal with completes the list of phenomena describing classical electrodynamics; that is *induction*, also called *Faraday's Law (Michael Faraday (1791 – 1867))*.

When we place a loop of a copper wire in a magnetic field (Fig. 3.6) and we change the magnetic field in the loop, in some way a current is created in the wire. A current, or an electric force leading to a current, is induced in the loop. It doesn't matter how the magnetic field in the loop is changed, whether the field is switched off, or the loop is tilted so the magnetic field in the loop is changed, decreased or increased, or the loop is taken out of the magnetic field; in all cases a current is induced.

In case the loop is moved the induced current can also be explained by the Lorentz force.

By rotating the loop around an axis perpendicular to the direction of the magnetic field one half of the loop experiences the Lorentz force in one direction whereas the other half experiences the Lorentz force in the other direction; both forces produce current circulating in the same sense.

Considering the last case, taking the loop out of the magnetic field, we also can explain the current arising in the wire by the Lorentz force. By removing the loop a part of the wire is moved in the magnetic field so that the wire experiences the Lorentz force leading to a current which is not compensated by any current in the other oppositely directed part of the wire loop when this is already outside the magnetic field. So, part of the induction phenomenon can be explained by the Lorentz phenomenon.

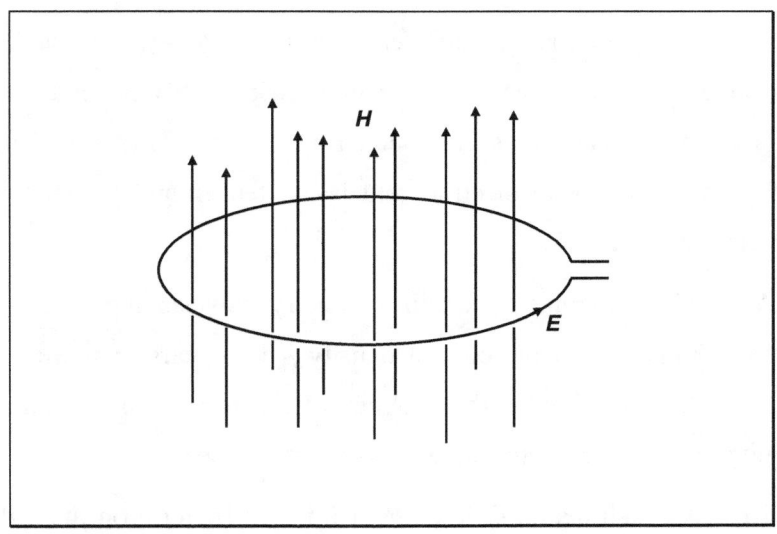

Fig. 3.6. Induction.

Around a changing magnetic field **H** an electric field **E** is developed. In a copper wire around the changing magnetic field this electric field causes a current.

The change of the magnetic field may be a real time-dependent change or a change due to a movement of the wire loop, for example due to rotation of the loop or due to motion of the loop out of the magnetic field.

We can explain that in a different - more general - way by saying that around a changing magnetic field an electric field is induced. And that is exactly what the second equation of Maxwell says. This equation tells us that an electric field is created around a changing magnetic field (Fig. 3.6).

You will recognize a certain symmetry between the first two equations (Fig. 3.1). If the current density j disappears then we can exchange the two fields, the magnetic and the electric without changing the structure of the equations except for the signs.

Now we shall deal a little closer with the first term on the right side of the first equation, the changing of the electric field. We shall see what is hidden in this term and try to understand what the symmetry means; we shall have a closer look at the electric field.

Suppose you have two parallel plates of metal with a small gap between them. Let one of the plates be negatively charged, i.e., having a surplus of electrons, and the other be positively charged, i.e. having a deficit of electrons. We can easily arrange these conditions by connecting the two plates to the terminals of a battery. Because of the gap there is no current. But the space between the plates is filled with an electric field. A single, freely floating electron in this space would move, would be repelled from the negative plate and attracted by the positive plate, i.e., it would move because of the influence of the electric field between the two plates. Thus we have an electric field between the two plates with the field lines going from one plate to the other.

If we now remove the battery and then connect the two plates by a copper wire, a current will flow through the wire from one plate to the other. According to the first Maxwell equation a magnetic field is created around the wire due to the second term on the right side, the term with the current density j (Ørsted!). What happens about the electric field between the plates?

Well, the current flows until the surplus of electrons on one plate has moved to the other plate and compensated the deficit there. Then nothing more happens. No current any more and the electric field has disappeared.

But during the time the current flows – the whole story takes place in a very short time – a magnetic field is created not only around the wire but also around the electric field between the plates, as if a current flowed across the gap between the plates and displaced the charges from one plate to the other. That means the changing electric field – the field is changed from some magnitude to zero – created a magnetic field.

The same happens when we recharge the plates with the battery. A current flows, the battery pumps electrons from one plate to the other, again generating a disequilibrium of electrons and an electric field between the plates. A magnetic field is created around the connecting wires to the battery and around the electric field between the plates – again as if a current flowed between the plates and displaced the electrons.

And that is again exactly what the first equation of Maxwell tells us. The current j, as well as the changing electric field E, creates a magnetic field H with the field lines going around the wire and around the changing electric field lines in circles. The direction of the magnetic field depends on whether the electric field is increasing or decreasing.

With the words increasing and decreasing I have already pointed to an important fact. The increase and decrease of the magnetic field takes some time; the changes don't occur instantaneously and don't take place in the entire space simultaneously. When the current is switched on, the development of the magnetic field starts directly at the copper wire and then expands. If the magnetic field arose abruptly in the whole space, then, according to the second equation, an infinitely high electric field would be created for a brief time, and that is not possible because it would require infinite energy. Hence, the magnetic field rises slowly – still very quickly but not instantaneously. Due to the negative sign in the second equation, the electric field that is induced by the increasing magnetic field has a direction opposite to the electric field produced by the current charging the plates. That means that the current in the wire is retarded, reduced by a so-called *self induction*, that is a negative feedback of the rising magnetic field on the very current which creates the magnetic field. Known as *Lenz's Law*, it acts as a brake on the current, forcing it to develop slowly, *(Heinrich Lenz (1804 – 1865))*. Slowly means for us still very quickly but not instantaneously.

It is also not possible for the induced magnetic field to appear simultaneously in the whole space because in this case the induced electric field would not fall off with increasing distance from the wire. In a very big loop parallel to the wire that would imply a very big change of the magnetic field in the loop and thus a very large induced electric field. Thus the magnetic field can only propagate from the wire with a finite velocity.

If we now add up all this, we shall see that something very strange will happen. When an electric field is changing, a magnetic field is induced around the electric field (the first equation). If the electric field is oscillating, that means changing its direction continuously, then the magnetic field is also oscillating, and with the same frequency. Around this oscillating magnetic field again an electric field is induced (the second equation) which then oscillates, and so on. A wave of oscillating electric and magnetic fields arises propagating from the first oscillating electric field throughout the space – an *electromagnetic wave* (Fig. 3.7).

The frequency of all these oscillations is equal to the frequency of the primary oscillator, which is just an antenna radiating the electromagnetic wave. With a little mathematics which I shall not explain here – that is just boring and only makes us tired – one can calculate that the velocity of propagation is the coefficient c in the first two equations.

The propagation velocity is the well-known velocity of light $c = 300,000$ km/sec.

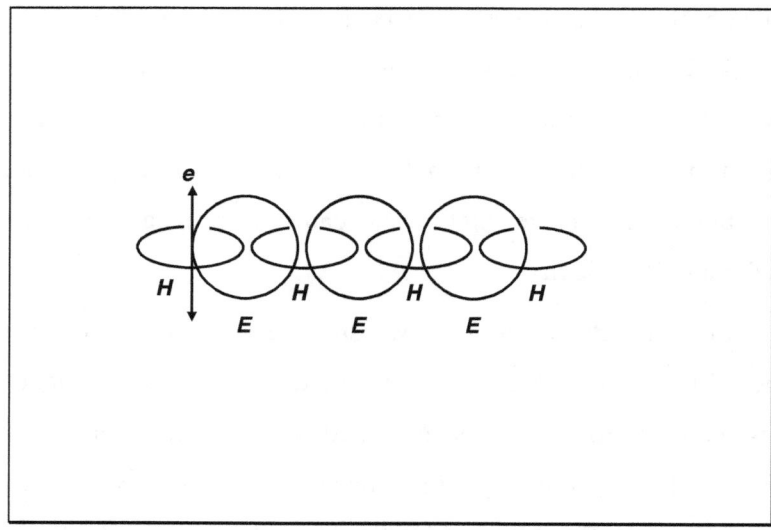

Fig. 3.7. The electromagnetic wave.

To the left in the figure an electric field is oscillating due to an alternating current of electrons moving up and down in an antenna.

Proceeding from left to right we recognize around the antenna a magnetic field oscillating with the same frequency. Around this magnetic field is an electric field with the same frequency and so on. A wave of oscillating alternating electric and magnetic fields develops – an electromagnetic wave is emitted from the antenna.

Here only the propagation from left to right is shown; in fact the wave propagates with maximum strength in all directions perpendicularly to the antenna, not only in the plane of the paper, but in all other directions.

This velocity is the same for all frequencies. That is also the velocity with which the circular magnetic field lines surrounding the wire will propagate out if the current (in Fig. 3.3) is switched on.

Maxwell's equations describe the propagation of electromagnetic waves over the entire spectrum of wavelengths from long-length radio waves, through infrared waves, visible light, ultra-violet light, to the shortest-length gamma radiation.

All these waves have the same velocity but vastly different wavelengths. The light of the sun, familiar to ancient man, and the radio signals between our cell phones are both electromagnetic waves in the framework of classical electrodynamics and described by Maxwell's equations. We sense the world contained in these equations, the whole world of modern electricity and electronic communication.

We shall now take a look at what classical electrodynamics says to a single electron, how it makes that electron behave.

Suppose a single electron is passing us; what do we notice? Well, when it comes closer we notice an increase in the electric field of which the electron is the source, in perfect accord with the third Maxwell equation. When the electron is passing we measure the maximum of the field, and when it is departing the field disappears again. That means, we see a short increase of an electric field to a maximum and then a decrease back to zero. Therefore, a magnetic field should be induced which also is changing. Following this further we should see an electromagnetic wave; the electron should cause a flash by passing, a wave with a frequency which depends on the velocity of the passing electron. But since the electron doesn't know where we are standing with our measuring instruments it should radiate waves from its entire track; the track should glow.

But we know from experiment, that is not the case; the electron actually moves inconspicuously. Also the electron wouldn't be able to emit light all the time without becoming exhausted and disappearing. The energy has to come from somewhere, for example, by accelerating from an external force or by decelerating.

So, what does happen?

In fact, the electron emits light or energy as electromagnetic waves all the time. But the wave it is emitting at one moment interferes with the wave it emits in the next moment, a little bit later. And if the next moment just corresponds to half a wavelength then these two waves annihilate each other by interference. Now, for every point on the track you will find a later point on the track from which a wave is emitted that annihilates the previous wave. The end result is that the electron doesn't emit at all as long as it moves steadily. Alternatively, you can say the electron emits photons all the time and immediately reabsorbs them.

But if the path of the electron is disturbed, if it is deflected by another charge, then the pattern of the interference with total annihilation is also disturbed, and some wave will be left for emission. As we saw on the 1st day, when the movement is disturbed by deflection or by stopping at some obstacle, the electron emits electromagnetic waves, so called "Bremsstrahlung". The emitted radiation can be measured, for example at the TV screen where the electrons are stopped. Also X-rays are produced the same way; they are "Bremsstrahlung", too. Here the electrons behave classically; they conform well to the laws of classical electrodynamics.

For the same reason they also should emit light when they revolve around the atomic nucleus. On this orbit they continuously change the direction of their movement – so we are again amidst the difficulties from the 1st day. Here we have hit upon the limit of classical electrodynamics.

What I wanted to show on this day was how it is possible to develop an "understanding" of many confusing phenomena with the aid of a few given facts. The key idea is to keep in mind that what nature presents to us is often not possible to understand in the sense of logically deducing it per se. Instead, we just have to accept the facts. We are not able to "understand" why a magnetic field is formed around a current, or why electric charges exist at all; we can only accept these facts.

Of course, we can explain the existence of elementary particles (as, for example, the electron) in the framework of an improved theory. But then we have only transferred the problem of understanding on to the next more general level. Even in such a broader theory there are things we can only accept but not really understand. The art is to uncover and to distinguish what nature presents us with and what we can conclude in order to base a theory upon as few assumptions as possible – that gives us the feeling of understanding.

Also there is nothing mystical or supernatural in nature – which doesn't mean that we are able with our brain to understand all. But there is so much wonderful – open or concealed – in nature that fills us with awe.

4th day, antimatter, general relativity, and gravitation.

We shall now come back to the electron. The electron in the old "classical" concept exhibited velocities far above the velocity of light, which meant that something was basically wrong with the classical concept. The idea of the electron as a small spinning ball had to be abandoned – but what was it then?

The electron could not be described with the tools of both classical electrodynamics and the theory of relativity without severe contradictions. While its behavior could be described by means of quantum theory, nothing was known about its structure. And then in addition to that deficiency came this above-mentioned (the 2nd day) flaw concerning the Schrödinger equation.

As I promised earlier (the 2nd day), we now come to an experimental example of the principle of correspondence. We had to take our leave from the idea of the small spinning ball. The spin, interpreted as a mechanical angular momentum, led to difficulties; the electron rotated too fast. Nevertheless, Albert Einstein and *Wander Johannes de Haas (1878 – 1960)* got the idea to adjust the magnetic moments of all the electrons in the atoms of a piece of iron with the help of an electromagnet. Normally, the directions of the magnetic moments are randomly distributed. But if the iron is magnetised, they will point more-or-less in a common direction. And that was what Einstein and de Haas wanted to achieve. In this case the mechanical angular momenta of the spinning electrons should also adjust to a common direction, and the piece of iron should turn. Indeed, when the electromagnet was energized, the iron piece did turn.

But at the same time the cable channel caught fire because of the strong current pulse. The experiment had two consequences: First, Einstein and de Haas became famous, and second, they were banned from doing any more experiments.

The experiment was later repeated by others with more precautions. Although the spin of a single electron cannot simply be described as a mechanical angular momentum, the sum of all the spins in the piece of iron does result in a classical mechanical angular momentum. That means there exists a classical correspondence to quantum mechanical spin.

About the same time that Schrödinger set up his equation with which he described the electron as a wave, Heisenberg developed a theory in which the jumps of the electrons in an atom were arranged in a scheme or spreadsheet. The columns were ordered according to the possible initial states and the rows according to the final states. In the rubrics were the so-called transition probabilities, the probabilities that such jumps could occur. Such a spreadsheet is called a matrix – and therefore the whole theory was called *matrix mechanics*.

Schrödinger's wave mechanics and *Heisenberg's matrix mechanics* at first seemed to be two totally different theories that were equally adequate in describing the atomic world, but we didn't know why. *Paul A.M. Dirac (1902 – 1984)* later succeeded in demonstrating that both theories should give identical results because they are simply two different representations of one common theory.

This common theory works with a mathematical formalism developed by the mathematician *David Hilbert (1862 – 1943)* at the beginning of the 20th century. He did this work without knowing it would find an application in physics.

When light emitted by excited atoms is sent through a glass prism, one sees a spectrum of colorful bands that arise from the jumping electrons. From the bands one can deduce how the electrons are arranged in orbits around the nucleus of the atom. As gravitation helps us correlate parameters of the planets around the sun, quantum numbers make it possible to keep order in the tiny world of the atom.

But questions were still open, for example the question of why the electrons do not all collect in the lowest orbit, the orbit closest to the nucleus with the lowest energy. As we saw on the 1st day, that would surely be the most stable state.

In order to describe this – on purpose I say describe, not understand, only describe – a *postulate* had to be established.

Well, what is a postulate? A postulate is just a statement of "that is how it is".

We all know several postulates, only we have gotten so used to them that we don't feel them as postulates. They have achieved a level of implicitness like, for example, the postulate: "Every cup we drop in the kitchen falls to the floor."

That is a postulate even if it doesn't sound very scientific. Newton expressed it in a different way: "Masses attract each other," and it is called Newton's law of gravity. In our special case in the kitchen, the earth attracts the cup with a certain force, or more completely, earth and cup attract each other. Because the cup is heavy the force could be called heaviness force, and that is what it is called in German: "Schwerkraft"; in English it's force of gravity. In the case of the cup this force is 300 grams (~ 2/3 lb) at least my letter balance shows this when I put the cup on it. Not only does the earth attract the cup with this force, the cup also attracts the earth with the same force. They attract each other equally.

The advantage of this general formulation of our cup postulate is that we realize that the same law that applies to our cup and earth applies to moon and both earth and the water in its oceans, thus causing the tides. That all can be explained with the Newton postulate. We have not explained in terms of understanding why the cup falls to the floor, but we have made some order in various phenomena.

Now back to electrons not collecting in the lowest orbit of an atom. That phenomenon is "explained" by the *Pauli principle*, a postulate formulated by *Wolfgang Pauli (1900 – 1958)*. That principle is to be differentiated from the Pauli effect, which is the breaking or falling down of equipment whenever that distinguished *theoretical* physicist entered a laboratory.

To understand the Pauli principle we have to review electron spin. This spin has a strange property. One might expect that all these small tops were distributed in the orbits so that the directions of their spin axes were randomly directed – in the classical picture, but that is not the case. In fact, if a magnetic field is applied to an atom, the electron spins become oriented so that they all point in the same direction as the field or in the opposite direction. Only these two states are possible.

The magnetic field need not be an external field. Each electron moving around the nucleus creates a magnetic field like the current in a coil, and often the atomic nucleus posses a magnetic field. From its movement around the nucleus the electron has a so-called angular momentum and an associated magnetic strength or moment. Actually there are two angular momenta, one from the spinning motion of the electron and one from its orbital motion, like the earth spinning around on its axis in 24 hours and orbiting around the sun in a year. Because the electron is electrically charged its two motions create two magnetic moments, a spin magnetic momentum and an orbital magnetic momentum.

Whether the magnetic field is applied from outside, or is created by the other electrons, or by the nucleus, or is the sum of all these fields, the spin of the electron is aligned either parallel or anti parallel to the combined resultant field. These two states are designated with quantum numbers. But because we only have two states they are numbered not with 1 and 2 but more practicably with **+** (plus) and **−** (minus) or with "*up*" and "*down*".

The Pauli principle now says that no two electrons in an atom can have all of their quantum numbers the same.

Therefore, in the lowest orbit there is only room for two electrons, one with the quantum number "up" and one with the quantum number "down". With this principle it was possible to describe the periodic system of the elements as displayed in the Mendeleev chart *(Dmitri Mendeleev (1834 – 1907))*. Chemistry became "understandable". Now it was clear why the alkali metals were monovalent. It was even possible to explain why iron, cobalt or nickel could be magnetic.

Because of the exclusion of two electrons in an orbit having the same quantum numbers, the Pauli principle is also called the *Pauli Exclusion Principle*. Also, the Pauli *Effect*, was renamed the Second Pauli Exclusion Principle, and Pauli was excluded from laboratories.

Our use of classical pictures may be very bewildering, but these pictures do show how the quantum numbers were developed. We use them as a memory support; they help us remember what the numbers mean. But we are not allowed to conclude anything; at least we have to be very careful about conclusions. Otherwise, we would very quickly get into troubles. Orbits, tops and similar things are just corresponding classical terms for the quantum mechanical terms used to describe the micro cosmos.

The trouble with the Schrödinger equation not being consistent with relativity theory, as mentioned above, created a task for many physicists. Pauli, Dirac, Gordon and others were searching for an equation, a description, that ought to fulfil several demands. Pauli set up an equation that was an extension of the Schrödinger equation. We have just seen that the spin of the electron has the peculiarity that it is aligned in a magnetic field only in two possible directions. This peculiarity was not considered in the original Schrödinger equation, but the Pauli equation took it into account. Since the Pauli equation was also not Lorentz invariant, adherents of relativity theory raised objections against the Pauli equation. The search had to go on.

Walter Gordon (1893 – 1939) and Dirac then set up an equation that apparently met all demands. The Schrödinger fans were content because all that was described by their equation could also be described by the Dirac equation. The Pauli fans were content because the spin of the electron was included. The relativists were content because the Dirac equation was Lorentz invariant. Such agreement reminds me of a parliament creating a law with which all parties could at least be content.

But then something very strange was detected in the Dirac equation. It appeared that the calculated energy of a particle could be negative. That, of course, was nonsense. *Negative energy* doesn't exist in physics, at least not in classical physics. The equation exhibits a solution for the square of the energy. Therefore, the energy can also be negative.

A lot with 20 m by 20 m side lengths has an area of 400 m^2. On this lot you can build a nice little house. But a lot with the side lengths minus 20 m by minus 20 m is also a lot with 400 m². A lot with negative side lengths! I have never seen such a lot.

Well, of course, one could just ignore the negative energy; we need only consider the positive energy as a solution of the equation. But a theory that describes more than nature allows is unaesthetic. You should not have to explain when the equation is valid and when it is not valid.

Dirac ignored all these difficulties and claimed the negative energy does exist in nature. He imagined the nothingness, the physical vacuum, to be like a great lake filled with electrons all having negative energy (Fig. 4.1). This lake, filled with electrons, normally doesn't appear. But at times it could happen that an electron in the lake absorbs energy from an electromagnetic radiation field and thereby accesses the world of positive energy. An electron suddenly appears. Thereby, a cavity is left in the lake. Where the electron had been a hole appears.

Relative to the charge of the filled lake, the charge of the lake with one electron missing, that is, the charge of the hole, appears with the opposite charge.

That means that the hole left by the electron in the lake has a positive charge. The hole appears as a particle with positive charge. Therefore, it was named the positron. The positron's existence was purely theoretical, invented only to explain the negative energy in the Dirac equation.

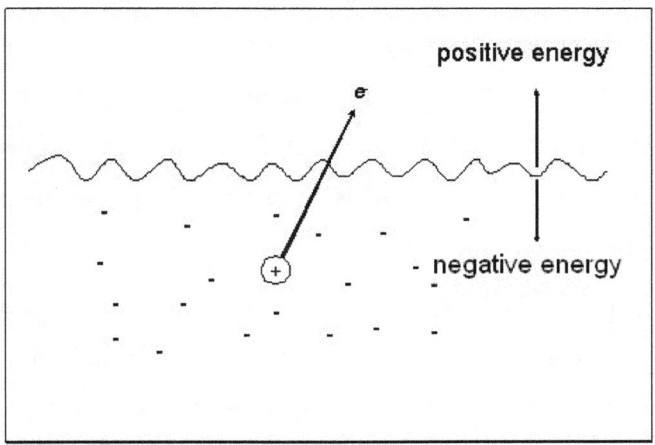

Fig.4.1. Dirac's electron sea.

Dirac imagined the vacuum to be like a huge lake, a sea filled with electrons of negative energy not perceivable by us.

But if an electron absorbs energy enough to get into the world of positive energy, not only the electron appears but also a positron – that is the positive hole left in the negative sea.

Whenever a theoretician anticipates something, experimenters start the hunt, either to find what the theoretician predicts or to prove that it is nonsense. Generally, they don't do that to bother the theoreticians but to test the theory. The more a theory is scrutinized the stronger will the theory come out from the test – if it is not disproved. Certainly, ambition is also a motivation; physicists are also just human beings.

And experimenters led by *Carl Anderson (1905 – 1990)* found the positron. In an electromagnetic field a positron and an electron suddenly could emerge. The electron from the sea with negative energy had caught from the field the appropriate energy it needed to make the leap into the real world with positive energy. Simultaneously, the positron, the hole left in the sea, appeared with its positive charge. An electron-positron pair had been created.

The positron also had the same mass as the electron, but it was no normal matter, it was *antimatter*, it was the antiparticle of the electron. It also had a spin, exactly like the electron, the same value and it had the analogous magnetic moment. So it was a small spinning ball made of antimatter. The electron had gotten a real twin sister.

But which is matter and which is antimatter? That is a mere case of definition. We could have called our normal matter antimatter. But that should have been done by our old Greek philosophers, and then we would have had to wait three millennia for the discovery of matter.

More in the theory was verified. The absorbed energy E from the radiation field was in agreement with the previously mentioned

Einstein equation $E = mc^2$ with m as the combined mass of the two particles.

Also, the reverse process appeared possible. The electron, the just-created electron or another electron, could fall into the hole in the sea. Then both the electron and the hole, the positron, would disappear and the equivalent energy appears as electromagnetic radiation, as a gamma-ray quantum, a photon.

In fact, two gamma-ray quanta appear, going out back-to-back so that linear momentum is conserved.

That is the so-called *electron-positron annihilation process*; the previous process is the *pair production process*. In these processes energy and matter are converted into each other.

In classical physics we have the matter conservation law, which says that matter cannot be destroyed or created. In the same way we have the energy conservation law, which says that energy cannot be destroyed or created. In modern physics, which of course includes relativity, we have to unite these two laws into one more general law, which says that energy and mass together cannot be destroyed or created. The sum of matter and energy remains constant in a closed system. But matter can be changed into energy and energy can be changed into matter.

The positron is as awkward as the electron against attempts for explanations. The small spinning ball of positive charge leads to the same nonsense, as does the ball of negative charge. But both, electron and positron, belong together. Whenever we try to catch the electron, the positron also appears – and thereby automatically also the photon. In a suitable theory this trio – electron, positron, and photon – has to appear together, but how is still incomprehensible.

In 1955 antiprotons were produced in the laboratory, and subsequently other antiparticles were also produced. This is strong evidence that for every particle there is an antiparticle. These antiparticles yet have no long lifetime because they very quickly come into contact with normal matter that is everywhere so that in a mutual annihilation both are turned into energy. As with the electron and positron, that energy is often in the form of electromagnetic waves, but with the annihilation of heavier particles not all of the mass is converted into energy. From proton-antiproton annihilation, for example, sometimes a pair of high-velocity pi mesons emerges. The mass of a pi meson is about 1/7 the mass of a proton. So most of the original mass has been destroyed and converted into energy of motion, the kinetic energy of the mesons instead of the energy of two gamma rays.

The whole universe consists primarily, if not entirely, of matter. At least it is not likely that somewhere a higher amount of antimatter has collected because at the locations where they would contact each other we would see the strong annihilation radiation. Astronomers haven't seen anything of that sort.

Probably in the big bang, when the universe was born, almost the same amount of matter and antimatter was created. Annihilation occurred in the first second after the big bang and created a huge flash.

The matter we have in the universe today is what was left over, a tiny excess of matter. The afterglow of this flash is seen now, after 13 billion years, as the so-called *cosmic background radiation*, which has cooled down almost to the absolute zero point. I tell this only to show how the physics of the micro cosmos plays a role in the state of the macro cosmos.

Like normal matter, anti matter is subject to the force of gravity, i.e., a stone of anti matter would fall to the ground like a normal stone – if it wouldn't be destroyed by annihilation with the atmosphere before reaching the ground. A lump of anti sugar would, in a fraction of a second, be turned into radiation of enormous energy, of unimaginable lightning. A peace of anti sugar tossed into the coffee would give a real surprise.

But what anti matter "really" is we still don't know – as we don't know what matter "really" is.

Even the photon, this curious particle of the electromagnetic wave field, has an anti particle. That is also a photon. The photon is not matter; it has no mass. Therefore, the anti photon also is not matter and has no mass. Hence, the photon and the anti photon are not distinguishable. They are both photons, or we can say the photon is its own anti photon. When they collide nothing happens, they are already pure energy.

The photon also shows another peculiarity. It is moving with the speed of light. Therefore, according to the theory of relativity, a clock moving with a photon goes infinitely slowly; it is standing still. That also means that the photon in a way stays young forever. Photons that were emitted billions of years ago, according to our clock, are as young as they were on their first day. Old photons don't exist. The cosmic background radiation, the flash from the creation of the world, has not grown older. What is hiding behind that enigma will probably remain a mystery.

When photons arrive from distant stars they tell us, "we were just born."

"But how can that be?" we may ask. "You have been travelling from stars and galaxies thousand of billions of kilometres away; and Einstein says you cannot go faster then 300,000 km/sec. There must be something wrong. According to my calculation you have been on the voyage for millions of years."

But the photons chant: "No, indeed we were just created; we have been on our trip for less than a second. For us the universe is only a tiny space. Effectively, we are everywhere simultaneously."

To understand this we again have to take our place at the roadside and observe fast cars.

Such cars are normally about 4 m long from the front bumper to the rear bumper. If we measure the length of a very fast car we notice that it is only 3.5 m long, provided that it has a speed of 145,000 km/sec, i.e., about half the light velocity. So not only do moving clocks go more slowly, but moving rulers also shrink. We see a length contraction of objects in the direction they move.

Now we also recognize the difference between the old Galilei transformation and the new Lorentz transformation. In the Galilei transformation we have space with its three dimensions of length, width, and height. And we have time like a universal clock ticking independently of our movement in space. That is what we are used to when we are moving with civic velocities in road traffic.

But when we move with very high velocities, we see that space and time fuse into one four dimensional unit.

How can we understand that? Well, it is not so difficult to understand, only we can't imagine it because our brain has been developed in a three dimensional world. Our ancestors didn't have to move with light velocity to hunt mammoths or to pick berries. It all went very unhurriedly – compared to our space shuttles today. Such a satellite is moving around the globe in 1½ hours.

In our everyday world are the three dimensions length, width, and height melted together into one unit, distance.

If we place a cube on the table, we can measure its height by placing a book on top of the cube and using a ruler to measure the distance between the book and the table. Why that complicated? Well, we want to measure the height of the cube, defined as the distance between its lowest point and its highest point, independently of how the cube is oriented in space. We start with the cube on the table with four sides vertical, but we shall change that. If we tilt the cube, the book is lying on the upper edge of the cube and the height of the cube, which is the distance between table and book, has increased. We only have to take care that the book is always parallel to the table.

Now we see that the height of the cube consists partly of the original height prior to the tilting and partly of the width of the cube (if we have tilted it on the length edge). That means the height of the cube, tilted or standing on the table, depends on how we tilt or rotate the cube in space. The three dimensions are one unit in the meaning that each of them consists of all three depending on how the cube is rotated.

In the four-dimensional world the changing of the velocity is equivalent to a rotation. The four-dimensional world of the passing cars – their length, width, height, and time – is tilted against our four-dimensional world. The time we measure on a clock in the moving car consists not only of our time but also of our spatial dimensions. The term "spatial dimensions" is just an intellectual shorthand for length, width, and height. Such terms are preferred because being shorter, you don't need to write so much, and also because using them shows how clever you are.

It is difficult to imagine such a four-dimensional world; and that is not necessary at all; you can instead consider the whole story as a mathematical trick with which we describe physical relations.

And we can help the imagination by stepping down two dimensions. Imagine two rulers lying on the table, not parallel but just how they would happen to lie if you threw them on the table as in the game of pick-up sticks.

These are two one-dimensional worlds in which the rulers "know" only one dimension, the length. On each of the rulers is sitting a fly, a one-dimensional fly knowing only the one dimension, the length. Therefore, the fly only can crawl back and forth on the ruler. Now these two flies start a conversation, a conversation about the length of the centimeter segments on their rulers (Fig. 4.2).

The first fly states: "Your centimeters are shorter than mine. I clearly can see that when I project my segments onto yours."

The other fly argues: "I see with the same method that your segments are shorter than mine."

They are both right. But they cannot figure out the reason because they can only run along the rulers. But from our three-dimensional world we can easily see that they indeed both are right and that their contradicting conclusions are only the result of how they look at their neighbour's ruler. The rulers are not parallel but tilted so that the projections of the segments are shorter.

That is so easy if you can see three dimensionally, and relativity theory would be so easy if we were able to see four dimensionally.

Since photons move with the speed of light, their universe shrinks to one point, and the clocks in our world stand still for them. That's how they can stay young forever in our world and how they can be present everywhere simultaneously.

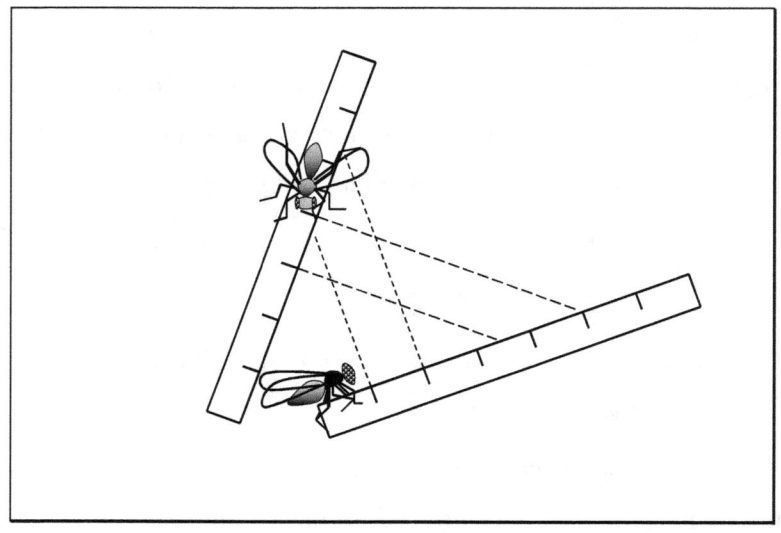

Fig.4.2. Shrinking of measuring rods.

Each of the two flies sees the other ruler with a certain angle. Therefore, they both see the centimeter segments on the other ruler shorter than their own centimeter segments.

Maybe somewhere here lies the key for an understanding of the collapse of the wave function I told you about on the 2nd day.

We saw that time is relative and length is relative. Now for a third peculiarity of the theory of relativity. By driving with our speedy car on a scale – you know these big scales used for trucks – we discover that our car has become not only shorter, but also heavier. The faster the car is, the heavier it is. The old law about mass conservation has to be revised because mass is dependent on velocity. Therefore, we have to distinguish between mass and *rest mass*.

Rest mass is the mass of the car when it is at zero velocity. That is the mass the driver, who is moving with the same speed as the car, measures. Relative to him the car is not moving; it has zero velocity.

When the velocity comes close to the light velocity the mass gets very high. At light velocity the mass is infinite. That is the reason why no object can ever reach this velocity. The closer an object comes to the light velocity the heavier it becomes and the stronger the force must be in order to accelerate it further. Rocket propulsion can be as strong as you want, and you can come closer and closer to the light velocity, but never reach it.

Only photons travel at light velocity. The earlier statement that photons have no mass refers to rest mass; they have no rest mass. Photons can possess a mass greater than zero, but not an infinitely high mass, and they can collide with electrons like billiard balls collide. That would not be possible without any mass or momentum. The momentum of the photons we will deal with later.

But what does the scale with the moving car show? We have to take a closer look at that. If it is a digital scale then it must show a certain number for the weight of the car, and both we and the driver must see that number. That's pretty obvious, isn't it?

If we were standing on the roadside we would see the scale showing the car to weigh more than its rest mass. If, for example, it is moving at 260,000 km/sec, it has twice the weight of a resting car. But the driver states that the scale shows only the weight of the rest mass. And he is also right. Is that now a contradiction?

Some people use such contradictions to prove that relativity theory can't be true. This contradiction – we shall pretty soon see that it is only an apparent contradiction – is an example of similar contradictions detected with such thought experiments. A thought experiment (in German: *"Gedankenexperiment")* is an experiment you only can do in your mind. You can only pretend a car driving with velocities close to light velocity; no car drives at 87% of light velocity.

Physicists try with such *gedanken* experiments to discover contradictions in order to lead the theory *ad absurdum*. The discussion around such *gedanken* experiments often gives deep insight into the nature of physical problems. We shall later (the 5th day) hear about another famous thought experiment and a connected paradox.

But now back to our small paradox about the scale. What does the scale show? Is this contradiction so serious that it could lead relativity theory *ad absurdum*? In order to clarify this we have to look very carefully at what the scale records. The scale records the weight of the car. The weight of the car is the attractive force between earth and car. Relative to us the car has twice the mass it would have if it were not moving. Therefore, the attractive force between earth and moving car is twice as high. The scale shows *us* this relativistic increase in the mass of the car.

But the car driver sees the scale showing the rest mass of the car, which is only half of the moving mass. However, to the driver the earth is moving backwards at 260,000 km/sec. That means the earth has, relative to the car driver, a mass that is twice as high as the resting earth. Therefore, the attractive force between the resting car and the moving earth is twice as high – and that is what the scale shows him. In both cases the scale shows the same, namely the twofold attractive force – for us because the car has the twofold mass, and for the driver because the whole earth has the twofold mass. Thus the contradiction is clarified and relativity theory is saved.

Unfortunately, not yet completely.

Let us consider the much simpler case of the car standing still on the scale. Then the scale shows the weight of the car, say 1.5 tons. That's what the car weighs. That is the attractive force between the mass of the earth and the mass of the car. Now let us enter a space shuttle and fly at 87% of light velocity relative to the earth. The mass of the earth then is twice as high, and the mass of the car is also twice as high. That means the scale should show an attraction force four times as high, i.e., it should show a weight of 6 tons. This contradiction cannot so easily be discussed away. The observer on the earth sees the number 1.5 on the digital scale, and we in the spacecraft see the number 6 on it – if we have got the relativity theory right. So, what number is really there?

In order to resolve this paradox it is necessary to abandon Newton's conception about mass and gravitation and to develop a completely new conception about gravitation. That is how the general theory of relativity arose, essentially from the cogitation of one single physicist: Albert Einstein.

Let us have a look at the relativity theory we have learned up till now. It strikes us that we have been dealing only with uniform velocities and that we have been dealing with gravity only marginally. This theory therefore was called the *"special" relativity* theory to distinguish it from the theory dealing with accelerations and gravity, the *"general" relativity* theory. The basic idea of the special relativity theory is the relativity of velocity. There exists no absolute reference frame in which an absolute velocity can be defined. We can only state relative velocities. If we are sitting in a train wagon with all the windows closed and covered without any connection to the world outside, then we cannot decide whether the train is standing or moving with a constant velocity. A standing train car that is shaking or vibrating is not necessarily moving anywhere. There is no experiment, either mechanical (for example by throwing balls) or optical (for example by measuring the light velocity), with which we can detect whether we are standing or moving. In fact, the question itself is senseless because of the relativity of velocity.

A similar basic idea we find in general relativity theory. In an elevator we feel lighter during the downward acceleration. The greater the acceleration, the lighter we feel. If the cable broke (no worry - *gedanken* experiment), we would feel weightless as our downward speed increased by 10 meters per second every second (abbreviated: m/sec^2) due to earth's attraction – i.e., after 1 second the velocity is 10 m/sec., after 2 second it is 20 m/sec., after 3 seconds 30 m/sec. and so on...

Sitting in a spacecraft orbiting the earth we are falling freely all the time. Because the curvature of our orbit matches the curvature of the earth, we never fall "down," although we are continuously accelerating. We don't feel gravity, the earth's attraction, and we are weightless. We are not flying in a gravity-free space, we are falling freely subject to the usual gravitational force; these are two different things. All items in the spacecraft are falling with us and are also weightless. And, because they all are falling with the same velocity, especially the spacecraft itself and all parts of our own body, we don't feel any gravitation. If all the windows are closed and covered so that we have no information from outside, then we do not have the possibility to determine whether we are far away from any planet or star, i.e., in the empty space between the stars, in the real gravity-free space, or if we are moving in a circle in the gravitational field of the earth. There is no experiment, either mechanical or optical, with which we could diagnose our situation. This inability to distinguish between gravity and acceleration is what is called *Einstein's Equivalence Principle*.

When we are standing on the ground with our spacecraft before launching, then in fact, we feel the attraction of the earth. We feel the same force way out in the empty space when we ignite our rockets and accelerate the craft at the special rate of 10 m/sec^2. In this case, again with all windows closed, we do not have the possibility to determine whether we are influenced by the gravitational field of the earth or are flying in the gravity-free space under the influence of the rocket's acceleration.

The idea of general relativity theory is that gravitation is the same as acceleration. In the acceleration view the floor we are standing on is moving up with increasing velocity, with an acceleration of 10 m/sec^2. That sounds crazy and impossible. And of course it is not possible in our three dimensional space. In this case the earth should swell, blow up like a balloon, and that is indeed not the case. But when we are located in four-dimensional space-time we can bend the space so that a movement along the time axis is the same as acceleration in space. Also that sounds crazy but it is not impossible. In four dimensional space-time we are moving all the time. Sitting quietly in a chair we are moving along the time axis, from yesterday to today and further to tomorrow even if we don't move at all during these three days.

Gravitation has become a case of geometry. The earth, or every mass, bends the space around it so that we feel acceleration, even if we are only moving along the time axis, i.e., sitting in a chair. The same applies if instead we are moving in space on one of the well-known ellipses around the earth, as already realized by *Johannes Kepler (1571 – 1628)*.

We are now coming back to our car on the scale. From the point of view of general relativity theory, the 1.5 tons of the car at rest on the scale are the result of the car's (and scale's) movement along the time axis and of the curvature of space-time caused by the mass of the earth. Observed from our spacecraft both earth and the car are moving at 87% of light velocity. That means their masses are doubled, both of them, which would cause the scale to read 6 tons. However, from the special relativity theory we in the spacecraft know that time on earth is going more slowly, by a factor of 2, than is time in our spacecraft time, i.e., the car is moving more slowly on the time axis – observed from the spacecraft. Therefore, the acceleration appearing as gravitation is smaller, only 2.5 m/sec^2 (because of the square of time), which is just so much smaller that the doubling of the two masses, the mass of the earth and that of the car is compensated. Hence, the mass scale also reads 1.5 tons for us in the spacecraft. These *gedanken* experiments have indeed been helpful and both cheaper and easier to perform than real experiments.

To see that, we had to revolutionize our view of gravitation as Newton had formed it. Newton had the idea of an attractive force between two massive bodies acting through empty space. In the idea of Einstein there is no remote action of attraction, there is only the warped space around the mass and the movement in the space-time continuum. Gravitation has become mere geometry. Of course, Newton's formulae are not false as long as we stay with velocities small compared to light velocity. If we calculate the mutual influence of earth and moon taking into account the warping of the space, we achieve the same formulae that Newton and Kepler developed for velocities small compared to light velocity. But the physical concepts are completely different.

5th day, polarization, radioactivity, and EPR.

Now watch out! The path is getting rockier and steeper. The term *polarization* was generally introduced to indicate that something has a certain direction – politically, as the direction between two extreme opinions, physically, as the direction between up and down, for example, or between left and right as well. Here we shall deal with the polarization of oscillations, of waves.

The well-tried old example is the jump rope children play with. We attach one end of the rope to a doorknob and then with the other end try to swing the rope. If a beginner tries to swing such a rope it mostly ends up in an unpolarized disorder. But after some exercise he will succeed in swinging the rope nicely in waves going up and down. Now the rope presents a vertically polarized wave. By swinging the rope from left to right instead of up and down we can produce a horizontally polarized wave. For our granddaughter to jump into the rope we have to let the rope circulate so that the middle of the rope makes a nice circular motion. Now we have a circularly polarized wave. Depending on which way we turn the rope we produce a right-handed or a left-handed circular polarization. Which one of them is right-handed or left-handed is mere definition. The best orientation for a definition is the corkscrew because the corkscrew is internationally known.

With all that oscillates, including electromagnetic waves, we can produce polarization. In a vertically polarized electromagnetic wave the electric field is oscillating up and down, which means the electric field of a passing wave is alternately pointing up and down. An electron exposed to such an electromagnetic wave is fiercely shaken, pulled up and down.

Sometimes an electron in its orbit around the nucleus is shaken so vehemently by an electromagnetic wave that it is thrown up to a higher orbit, to an orbit further away from the nucleus, and it stays there. The oscillating electromagnetic field disappears, the electron has taken all the energy, has absorbed it, just as the swinging rope collapses when the doorknob ruptures. In this case the doorknob has absorbed the energy of the oscillating rope. The doorknob is also thrown up to a higher level, probably on the top of the cupboard or in the lampshade.

Next time we try it – after having repaired the doorknob – with a circularly polarized jumping rope. The doorknob again is pulled out of the door and makes a huge circular move. We have given the doorknob an angular momentum, i.e., the circularly polarized wave has an angular momentum, which can be transferred to the body influenced by the wave.

Photons, the particles representing the electromagnetic wave, possess the properties of the wave, i.e., they are equipped with polarization and angular momentum whenever the wave possesses these features.

We shall now have a closer look at what happens if an electron and a positron collide and annihilate each other, the two masses being replaced by electromagnetic energy of an amount given by $E = mc^2$ where *m* is the sum of the two masses.

The positron has a magnetic field and a correlated spin – we know that already from the electron – defining a certain direction. The electron sees this direction and aligns its own magnetic moment, and necessarily its spin, according to this direction either up or down. We have seen that earlier. In the same way the positron aligns its spin according to the spin of the electron. Therefore, just before annihilation we have a system of an electron and a positron in which the spins are either parallel or antiparallel. This electron-positron structure even has a name; it is called *positronium* although it lives only a ten billionth of a second before it is destroyed.

We can also tell something about the electromagnetic wave emerging from the annihilation of the positronium.

We can always choose a steadily-moving coordinate system for our experiment so that the center-of-mass of the two particles, the electron and the positron, is at rest in our system, i.e., the total momentum is zero. That is like two billiard balls colliding with equal velocity from opposite directions. Each of the balls has a momentum – you remember that momentum is velocity of the ball multiplied by its mass. Since the two velocities have opposite signs, the sum of the momenta, the total momentum, is zero. You can verify this by hitting a third ball head on with the two balls simultaneously from each side. The third ball stays where it was, i.e., it does not obtain any momentum.

The electromagnetic wave, or the photon, coming from the annihilation of the positronium would have a momentum even though the total momentum was zero before the annihilation. Since annihilation cannot create momentum, a second photon is simultaneously created with the opposite momentum. Two photons must be created leaving the point of crash in opposite directions.

Before going on with the annihilation of the positronium we shall deal with a less heavy topic from classical physics – to some extent for a bit of relaxation, but also because we need it for our further understanding.

Each body has a *barycenter*. That is the point where the total mass of the body could be concentrated without affecting its movement through the universe if it were a planet, or through the kitchen if it were a cup. If the total mass of the body were concentrated at the barycenter, then this point would behave like the body itself in the gravity field of the earth. The attractive force of the earth effectively only "sees" the barycenter. It doesn't "see" whether it is an extended body, a coffee cup or only the barycenter of a coffee cup.

That is what is meant by the sentence "the force is acting on the center of mass, on the barycenter". In other words, the barycenter of the earth attracts the barycenter of the coffee cup.

Just as the barycenter of a single cup is also the barycenter of all the atoms of that cup, the barycenter of a collection of coffee cups is also the barycenter of all the cups in the collection. The barycenter of two cups can be found by connecting the two cups with a thin rod and with one finger finding the point on the rod where the two cups are just balanced. If the two cups have the same weight the barycenter will lay half way between the two cups.

When the waiter comes between tables and chairs with one hand under a tray filled with glasses, that hand is under the barycenter of all the glasses. He is carrying the barycenter of the glasses. If then a guest on his own takes a glass from the tray in order to help him, the barycenter of the remainder changes, and the waiter's hand is no longer under the barycenter. You can imagine what will happen even if you didn't know the physics.

If a grenade is launched, it will move on the well-known ballistic trajectory that comes from the combination of the initial velocity of the grenade and the velocity the grenade gets from its acceleration in the gravitational field of the earth. We all know that from throwing stones.

If the grenade explodes on its way, the fragments are scattered in the air. But the barycenter of all the fragments, which is also the barycenter of the former grenade, will continue on the trajectory as if nothing had happened, and all the fragments will scatter around in the landscape so that the barycenter "lands" exactly where the grenade would have landed if it had not exploded.

In the same way will the two photons created by the annihilation move apart so that their mutual barycenter stays where the positronium was before annihilation – if it was at rest. If the positronium was moving prior to annihilation, the barycenter will continue the movement the positronium would have had if it had not been destroyed – like the exploding grenade. In our coordinate system the photons go out back-to-back.

The whole story is summarized in the so-called law of conservation of momentum. Adding the momenta of the billiard balls prior to the impact gives the same result as adding the momenta after the impact. The total momentum is not changed, it is *conserved*. Admittedly, by adding the momenta we have to regard the directions of the velocities. How that is done is the business of vector mathematics.

A similar law of conservation also applies to angular momentum. We can vectorially add up the individual angular momenta to get a total angular momentum for which the conservation law applies. We have experienced that with the doorknob where the angular momentum of the jumping rope suddenly was transferred to the doorknob.

It should be noted that in discussing positronium we considered only the case where the two spins are antiparallel, i.e., where they compensate each other so that the total spin of the positronium is zero. We will leave it with that. In the electromagnetic wave the oscillating electric force is directed perpendicularly to the axis of the movement, i.e., to the direction of the photon's movement. It is similar to the rope's direction of oscillation being perpendicular to the line between our hand and the door knob – also if the rope is circulating.

Each of the two photons, flying diametrically apart, has an angular momentum, a spin, which means the electric force field of the associated electromagnetic wave is rotating around the direction of motion; they are *circularly polarized*, and both spins have to be either right handed or left handed, because the sum of both spins has to be zero, as was the spin of the positronium prior to annihilation. This deduction is in accord with the law of conservation of angular momentum. Since the photons are flying diametrically apart, their angular momenta will cancel only if they are either both right handed or both left handed. If you point a corkscrew in one direction and then in the opposite direction you immediately realize that the two rotating directions compensate each other.

Now we are approaching a famous gedanken-experiment, the Einstein, Podolsky, Rosen (EPR) Experiment of 1935.

The three physicists Albert Einstein, *Boris Podolsky (1896 – 1966)*, and *Nathan Rosen (1909 – 1995)* discussed an equivalent of the following arrangement:

Somewhere positronium is annihilated and two photons are emitted diametrically. The photons are caught in some detectors and their polarizations are measured.

Before we continue with the experiment we should know a little about the motives of EPR, why they made their thoughts about such an experiment. Einstein got the Nobel Prize because of his thoughts about impact experiments with electrons and photons, not because of relativity, although he is better known for relativity. These considerations showing that light consists of particles, photons, were the pioneering thoughts for the development of quantum theory. In spite of that, Einstein could never make friendship with this weird theory where causality is replaced by probability. He always tried to prove that quantum theory was nonsense, and the EPR gedanken-experiment was one of these attempts.

We have already seen that the uncertainty of prediction is one fundamental character of quantum theory. We can tell the future, what will happen and when it will happen, but only with a certain probability, not exactly. In scattering experiments with electrons we saw that we could calculate the probability that at a certain point on the screen an electron would appear. We cannot tell for sure that the electron will hit the point. This uncertainty, expressed by the Heisenberg uncertainty relation, is trailing through the whole theory.

You might ask if physicists have capitulated on finding exact answers to all questions. The result is indeed, yes, it is not possible to give exact answers. Not because we know too little about physics but because it is in principle not possible; nature itself doesn't know the exact answer.

Radioactive decay of an atom is a nice example. Therefore, we will take a look at that.

Our walk is more or less a random stray through the landscape of physics. Of course it would be possible to explore the landscape systematically. That would be convenient if you wanted to learn physics. Then you would start with the necessary mathematics and go logically through the chapters of physics. For this purpose there exist very good textbooks, although they can be rigorous and boring. If I would handle our walk like a textbook I would probably quickly be left alone on my walk. Therefore, I think it is nicer that together we have a look at the most exciting and most beautiful things without the need to understand it all. Anyway, with time you will come to the point where you see that you cannot understand it all, you can only enter the jungle and get deeper and deeper so that you know this or that trail, but what is behind the thick wall of trees and scrub will be forever concealed.

Well, now on to the decay of the atom. We already know that atoms sometimes decay, although the word "atom" comes from the ancient Greek and means indivisible. But we have before seen that names and descriptions do not always mean much. It is like when you see a butterfly and ask which butterfly it is. The intellectual zoologist tells you it is a brimstone butterfly with the Latin name such and such. Then we say, ah, now we know – but we only know the name, nothing else, and in the evening we have forgotten the name again.

The ancients did not know that some atoms are not indivisible. Radium atoms have a so-called half-life, a lifetime. Suddenly the atom disrupts, it disrupts into two parts and then it is no radium atom any more – like the positronium above, but the radium atom has a much longer life than the positronium. However, we don't know how long a particular radium atom lives. What we do know is that when we have a lot of radium atoms lying on our table half of them will have decayed after 1600 years. After another 1600 years again half of the rest will have decayed, and so on.

That is a very strange behaviour. We have no idea when any particular atom will decay. It might explode in the next minute or it might live for more than a hundred thousand years while the neighbour atoms on the table explode by and by. The atoms themselves don't know when they will decay, how long their individual lifetime is. But altogether they know that after 1600 years half of them have disappeared. That is a kind of collective knowledge of the radium atoms.

This decay behaviour is different from the mortality of animals or human beings. Imagine, for example, that we, maroon 1000 male persons on a lonely isle.

If these persons had a decay behaviour similar to the radium atoms but with a lifetime of 60 years, then after 60 years the number of men still alive on the island would be 500. After 120 years there would be 250 men, after 180 years there would be 125, and so on. After 600 years, 10 half-lives, perhaps one man would still be alive on the island. We immediately realize that radioactive half-life doesn't meet biological reality. The experience of human mortality is different.

Of course, there is no such collective knowledge of the radium atoms. All items decreasing by a certain percentage show the same behavior, for example money: if you use up ten percent every year of your asset you will realize a half-life of about seven years.

We don't know when a certain radium atom will decay, but one might think the reason could be that we know too little about the structure and the dynamics of the inside of that radium atom. The building blocks of the unstable radium nucleus, the protons and neutrons, are moving so vigorously that some of them could be catapulted out of the structure; the nucleus is in a way boiling. If we would know the momentary condition of the each component in that nucleus, then we might be able to precalculate its time of decay.

When a radium nucleus decays, it breaks down into two parts. One part is the nucleus of the element radon, a nucleus that is almost as heavy as the radium nucleus. The other part is a smaller nucleus, that of the element helium. In other words, the radium nucleus emits a helium nucleus, and a radon nucleus is left over. The radium nucleus consists of 226 particles, the helium nucleus of only 4 particles, and the left over radon nucleus of 222 particles.

So one might get the idea that if we only knew how the particles in the radium nucleus were moving and how they were combining, if we knew more about the dynamics of clumping inside the nucleus, then we could calculate when the helium lump, the four particle lump, would be built and ejected from the radium structure.

But that is exactly what we can't do. Because of the uncertainty relation it is impossible to state the velocity and the position at a certain moment of the particles inside the nucleus. That is not so because of lack in our experimental ability, that is so because the properties "where" and "how quickly" are not defined at all. These properties are created together with the two parts not until the moment of decay.

Therefore, the nucleus itself doesn't know when it will decay. Curious, but that is how it is.

We can only tell with a certain probability by when the radium nucleus will have decayed, for example, with 50% probability after 1600 years.

This kind of probability statement is an essential character of quantum physics. It is as if nature would throw the dice before it gives answers to our questions. And this was what Einstein never could accept, "God does not play dice with the universe!" is what he is reputed to have said. And that is the reason why Einstein together with Podolsky and Rosen conceived the gedanken experiment with positronium and the apparatus catching and detecting the two photons.

A final preliminary before we have a closer look at this EPR gedanken experiment is to look a little more at polarization. You remember circular polarization. That was the jumping rope where the girls are hopping when the rope makes circles. You can consider this circular polarization consisting of two linear polarization modes, one mode vertical and one mode horizontal. Imagine the rope is swinging vertically upwards, then one second later horizontally to the right, another second later vertically downward, still one more second later horizontally to the left, and after a final second vertically upwards again. The rope has done a full circle consisting of vertical and horizontal swings, which are phase shifted, that is, shifted in time. A vertical motion added to a horizontal motion 90° out of phase produces a circular motion.

In the same way it is possible to separate circularly polarized light into two linearly polarized waves, one horizontally, the other vertically.

If we send a circularly polarized photon into an optical filter, what comes through the filter will be a linearly polarized photon. We can't tell whether it will be vertically polarized or horizontally polarized, the probability for each is 50 %. The EPR-apparatus has an optical filter on each side of the positronium. When one side catches a photon from positronium decay, the other side catches the other photon. But now comes the strange thing about it. If we observe that the photon behind the filter on one side is vertically polarized, then we know that the photon on the other side will be observed to be horizontally polarized. That's what quantum theory says.

That doesn't sound strange at first, but it is strange because we now have the possibility to tell which polarization of two equally likely possibilities we will find behind the second filter. Suppose the second filter is far away and the second photon is still travelling. Then we know the polarization of this second photon although it is still on the way; it hasn't reached the filter yet. We know the polarization although the photon itself doesn't know its own polarization.

That is the famous EPR paradox that has been discussed a lot. With this paradox Einstein wanted to show the inconsistency of quantum theory. Either the polarization of the photon is somehow hidden before it arrives at the filter, which shouldn't be possible according to quantum physics, or there must be a signal received by the second photon from the first filter before it arrives at its filter telling it with which polarization it has to appear behind the filter. This signal has to be faster than the photon, i.e., faster than light, and that should not be possible according to relativity theory.

This contradiction only occurs in quantum theory, not in classical physics.

When you learn tennis you may start with a machine throwing tennis balls because the teacher is too busy with other beginners. Suppose you have a machine that throws two balls at the same time, a red ball in one direction and a blue ball in the opposite direction. If a red ball is coming to you, you will know that your partner at the other side will get a blue ball. You will know that even if your partner is far away and doesn't know which color his ball has; there is no contradiction.

But that is so because the colors of the balls are determined before they reach the players. The attributes of the balls are defined prior to the measurements. Not so in quantum physics. In the world of quantum physics the color of the ball is generated first at the moment the ball is caught. There is no possibility to define the color before it is observed. Quantum physical tennis balls have no color before they are caught.

Niels Bohr pointed that out. The color attribute is generated with the observation; prior to the measurement it is not defined. Such attributes are, for example, the attributes polarization and spin.

If that is the case – quantum physics states it and all our experiments show it – then there must be a coupling of some kind between the two photons enabling the exchange of signals with superluminal velocity, i.e., we cannot treat the two photons independently. They are somehow one object. Although they are far apart, one unit provides that the polarizations behind the two filters always have opposite directions, one vertical and the other horizontal.

Maybe this is associated with the strange thing we met the other day, that time is standing still for the photons, i.e., the second photon is not later at all than the first photon even if the second filter is far away. It only appears to us that they arrive at their filters at different times. In their own system they arrive at the filters at the same moment the positronium decays. For the photons it all happens at the same time.

In our everyday life we know subjects on the border between reality and illusiveness.

After the rain when the sun is warming our backs and we look at the dark rain cloud we see the rainbow, one of the most beautiful appearances nature offers.

This rainbow appears to us to be very real, but still it is there only when we look at it. When we look in another direction the rainbow disappears. You may say your neighbour also sees the rainbow even if you look somewhere else, and you can make photographs of it. On the picture you can see that the rainbow really is there.

The neighbour indeed sees the rainbow, but he only sees his own rainbow, not yours. You only see your own rainbow. And if the neighbour is looking somewhere else his rainbow disappears. Everybody only sees his own rainbow, and it disappears when he is looking somewhere else.

The camera also sees the rainbow, but only in the flash of the second the shutter is open. On the photo is only the image of what the camera sees in the split second the shutter is open. Before and after that there is no rainbow for the camera.

If you believed in the existence of the rainbow independently of your observation, some apparently sound and interesting questions could come to you. For example, how does the rainbow appear from behind?

You see, it is possible to put questions to nature that cannot be answered because they are senseless. Also your own mirror image, the image you look at every morning in your bathroom, is such a subject without reality – although you can get help by observing it from your mate or from the neighbour or even from a camera proving the existence of the mirror image. These mirror images have only a front side. They are hollow in the back like the elves in H.C. Andersen's fairy tales. The rainbow is real as an appearance. But as a subject it has no reality.

Even the shadow has more reality; the shadow is present even if nobody is observing it. We see that there exist different kinds and different nuances of reality.

6th day, the identity of the electron.

There have been a lot of philosophical discussions about the strange idea in quantum mechanics of probability replacing straightforward cause and effect.

We have a formalism, a method of calculation, with which we are able to calculate the probability of the occurrence of a particular event. This method works very well. In return, we have first to renounce the possibility of predicting – we can only tell with a certain likelihood what will happen – and secondly we have to get by without achieving understanding. We don't know why this formalism works so well – and that is what creates room for the imagination, for philosophical worldviews enlightening the background of the calculation – whether they do that is a mere matter of opinion.

But what is the benefit of this formalism if we cannot make predictions that are definite, but only have a certain probability? The result is just that our prediction is right with a certain probability, but also that our prediction is wrong with a certain probability – so what is it all good for? What is it that is working so well?

Help comes from the so-called collective knowledge, the law of high numbers. Suppose we have calculated that an electron will behave in a certain way with a probability of 0.5; then, in fact, we do not know how it will behave. But we can be sure that among one billion electrons 500 million of the electrons will behave as we have calculated – and that's not bad; we cannot expect more.

Whenever we find an electron with a velocity *v* at point *A* – as far as we can determine this within the limitations of the uncertainty relation, – we can calculate the probability of finding the electron after a time *t* at point *B*. The waves of probability of the electron propagating from point *A* superpose and concentrate at point *B* so that the probability is high to find the electron there.

With tennis balls it is the same. Only we don't realize that because the probability that the tennis ball will *not* be at point *B* is so negligibly small that it has not happened as long as mankind has played tennis – and for sure it will not happen as long as mankind will play tennis. All prospective tennis players can count on that.

With the small electron it is not that sure – the interference pattern of the probability waves shows that the probability of finding the electron at point *B* is not as high as it is for the tennis ball. And what makes the observation much more uncertain is that we don't know what the electron experiences between the two points. We only locate an electron, first at point *A* and then later at point *B*. We don't know if it is the same electron – we even don't know if we can talk about something like an identity of the electron.

It is very similar to waves on a water surface. We see the waves moving across the surface of the sea and we associate an identity with the waves. This wave, we say, is approaching our boat. But in fact the wave is only the movement up and down of the water on the surface. Does it make sense to say that the up and down movement of the water at the point A is the same as the up and down movement we see a little later at the point B?

The identity of the wave is questioned. At least we should know exactly what we mean by "the identity".

With the electrons it is the same. What we calculate are probability waves, and what we detect are light flashes on a screen or some other signal in a measuring device. And we associate electrons with these signals. But when we now don't know whether we detect particles or waves we also lose the certainty as to whether it makes sense to speak about an identity of the electron. At the point A something occurs that looks like an electron and at the point B a little later something occurs that looks exactly the same. But we don't know what happens in between. That is like playing tennis in the dark just equipped with a flashlight.

Even our own identity is not that certain. The atoms and molecule I consist of are surely not the same as they were seven years ago. Metabolism of an organism provides exchange of matter with the environment. Therefore, we say that our identity is not associated just with our body but more with the structures associated with our body, above all with the mental structure, with the processes in our brain – although our brain also is subjected to metabolic exchange with the surroundings. We experience the outlasting of these mental structures as memory. But we give a name to these structures and keep the name in a personnel record with a social security number. This file is normally not subjected to any exchange – it is eternal – and collecting dust.

Conceptions about the nature of probability waves are represented mainly by two different academies. One – around *Max Born (1909 – 1955)* – teaches the opinion that the probability waves are a mathematic construction, which states where the particle mainly is placed. The other school – the so-called school of Vienna around Erwin Schrödinger – has the opinion that the electron indeed is smeared out; the material wave represents this distribution. The electron in the surroundings of the nucleus is quasi everywhere at the same time.

7th day, polarized photons.

Polarized photons create mysteries. When talking about polarized photons we must remember that it is the electromagnetic wave which appears as a photon, that is polarized. Whenever such a photon hits a *polarization filter* it either goes through the filter or it is absorbed in the filter, depending on how the filter is oriented.

The photographer knows such filters. He puts them in front of his lens to avoid reflections from a window. As the light reflected from a window glass is polarized, the photographer has only to turn his filter into the right position to reject that reflected light. Such a polarization filter is thought of as a grid, a fence made of parallel bars. When you lead the well-known skipping rope through the grid you can demonstrate how the filter works. The grid is the filter and the swinging rope is the electromagnetic wave, the light. By swinging the rope in the same direction as the grid bars you can demonstrate how the wave goes through; the grid is no obstacle. But when you swing the rope perpendicular to the grid bars the waves are blocked by the grid and they don't go through the fence. The polarization filter works in a similar way. The molecules in such a filter are arranged so that they act on the electromagnetic wave like the bars act on the swinging rope. When you turn the filter so that the "bars" of the filter are vertical – vertical to the table where you have placed the filter – then only vertically polarized light passes the filter. The filter blocks horizontally polarized light. Light polarized in a plane between vertical and horizontal is equivalent to light with both a vertical and a horizontal component, and only the horizontal component is blocked.

The result is, of course, that from a stream of unpolarized photons, i.e., a stream of photons with all directions of polarization, the vertically polarized components pass the filter. Thus, with such a filter it is possible to generate polarized light from unpolarized light.

You can prove that by placing a second filter behind the vertical filter. If you turn this second filter also into a vertical direction, the photons will pass through it also; if you turn it horizontally all the photons are blocked; it is dark behind the second filter.

In summary, if the two filters are aligned parallel to each other, they let some light pass; but if they are aligned perpendicularly, they don't let any light through.

That seems understandable. You can imagine an entrance with two doors arranged serially. The first door is for women only. Then you have only women behind that door. If the second door is for men only, nobody can come into the building (Fig.7.1). That corresponds to the two filters aligned perpendicularly. If the second door is also for women only, all women can enter the building. That corresponds to the two filters aligned parallel.

Fig.7.1. Two polarization filters.

Two polarization filters arranged successively with perpendicular polarization directions are like two doors leading in series into a building. If the first door is for women only and the second door for men only, no one enters the building.

We shall now consider a special door with the feature that women can pass with 50% probability and men can pass with the same probability (Fig.7.2), i.e., if a man comes to this strange door he has a chance of 50% to get through – the same as for a woman. If a crowd of people comes to this door we will also have a crowd behind it, a smaller crowd because men and women have a chance of only 50% to pass through. Behind the door we have men and women again, but only half of the people who tried to pass the door; the other half have to stay outside.

That's how a polarization filter works if it is turned 45° relative to the vertical (and thus also to the horizontal). If vertically polarized light hits such a 45° filter, only half of the photons will pass the filter. Each photon has a chance of 50% to pass the filter – of course, the same is true for horizontally polarized light.

Fig.7.2. Slanted polarization filter.

A polarization filter that is turned 45° relative to the polarization direction of the incident light is like a door which lets women through with a chance of 50% and men also with a chance of 50%.

We shall see later that this door has a strange property. Although only women pass the door, men also appear on the other side. The door turns women into men.

Let's go back to the building with the two doors. Only women can pass the first door; only men can pass the second door – nobody gets into the building.

Now we shall put between those two doors the strange door, the door that lets both women and men through with a chance of 50%, so that we finally have three doors serially.

What will happen? Of course, still nobody can enter the building - we suppose.

Now we shall do the same with our polarization filters. We start with two filters – like the two-door-entrance to the building. First we turn the filters so that they are aligned parallel (Fig.7.3a) some light goes through the filters. Then we turn one of the filters so that they are aligned perpendicularly (Fig.7.3b); no light goes through the filters; it is dark behind the second filter. Finally, between the two filters, one of them aligned vertically and the other horizontally, we put the third filter aligned 45° relative to the two others – like the strange door. What will happen?

The first filter is passed only by vertically polarized light. The filter in between lets half of the vertically polarized photons pass. The last filter lets only horizontally polarized photons pass – thus, it should still be dark behind the three filters.

But it is *not* dark. Behind the third filter we have some photons, (Fig.7.3c). If you remove the 45° filter, it is dark again. If you slide it back in again, it is light behind the filters.

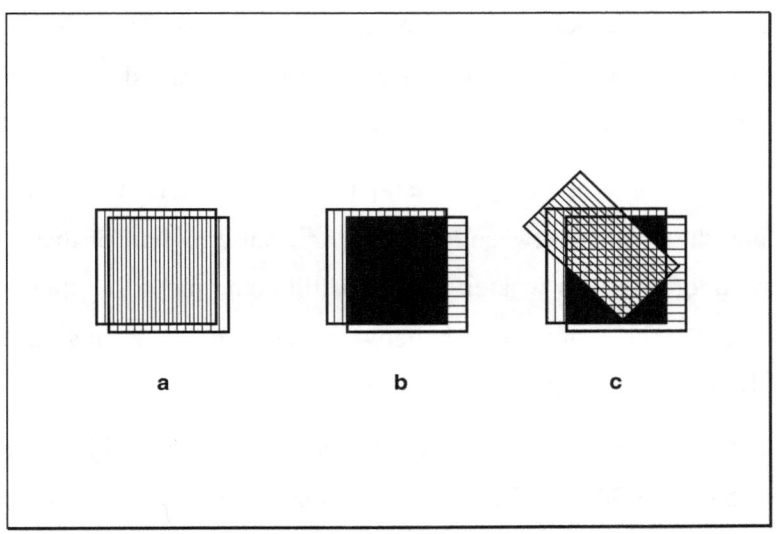

Fig. 7.3. Three polarization filters.

If the polarization directions of the two filters are arranged parallel, light can pass (a). If they are arranged perpendicularly no light can pass (b).

If a third filter is set between the two perpendicular filters and this third filter is turned 45° relativ e to the two other filters, then some light can pass all three filters (c).

Even though this 45° filter between the two others lets only a part of the photons pass, light comes through all three filters, although no light came through the two perpendicular filters.

Let us have a look again at the strange doors in our building. The effect of the third door we have put between the two doors is that behind the last door suddenly people appear, namely men.

The first door lets only women pass, the last door lets only men pass, and the door in between lets both pass, but only half of them. If the first door only lets women pass and still some men can enter the building, then the strange door in between must have an even stranger capacity. It turns women into men (Fig.7.2).

It is not just that the door in between lets women and men pass with a chance of 50%. In fact, a person – man or woman – who enters this door will be turned into a woman or into a man, each with a probability of 50%. In other words, a person who enters the door disappears and a person appears on the other side of the door – a woman or a man, each with a chance of 50%.

Hence, the identity of the person as a man or a woman is questioned. The same is the case with the photons passing the polarization filters. The photon hitting the 45° filter in between, whether it is horizontally or vertically polarized, leads to a photon appearing on the other side of the filter – a photon, which is horizontally or vertically, polarized, each with a probability of 50%. In fact, it is polarized in the direction of the 45° filter.

The difference between the filter in between and the other filters is only the orientation. That means a polarization filter does not simply let photons pass. Rather, it absorbs a photon and emits another photon on the other side. The emitted photons are polarized in the direction of the filter.

The photon hitting the filter loses its identity - if it has a meaning at all when speaking about the identity of photons - and a new photon appears on the other side of the filter. This conception about the photon and its polarization makes the strange behaviour of the filter understandable.

The properties of the polarization filter only seem paradoxical in the picture of light consisting of particles (photons). In the picture of the electromagnetic wave the polarization filter is quite understandable.

The oscillating electric field excites the molecules in the filter. The probability of excitation depends on the mutual orientation of the electric field vector and the direction in which the molecule can be excited, i.e., the direction in which electrons can move easily. The excited molecule radiates electromagnetic waves (light) like a small antenna. The polarization of the emitted waves is determined by the orientation of the oscillating electrons. In this picture no question of identity of the electromagnetic wave comes up.

But whenever we have to describe the properties of light with photons we have phenomena that are not understandable in the wave picture, and the question about the identity of single photons comes to us. So the photons are not particles in the common sense; they are just so existent that they can explain some of the phenomena but not existent enough for speaking about an identity. By trying to grasp the photon as a particle it disappears – like the electron, as we saw earlier.

We again see how the complementarity principle arises. We need both pictures – the electromagnetic wave and the photons, although they are contradictory – in order to describe the complete nature of light.

The picture of absorption and emission of electromagnetic waves explains not only the behaviour of the polarization filter; it explains the interaction between light and any object. For example, whenever light hits a surface of glass, the electric field vector excites the molecules in the glass. These excited molecules emit electromagnetic waves – light –, which again excite other molecules in the bulk of the glass, etc. Thus, electromagnetic waves propagate through the glass by repeated absorption and emission. The light that appears on the other side of the glass is emitted by the molecules on the opposite surface of the glass. That's why glass is transparent. Photons do not penetrate the glass. The photons on the other side are different from the photons on the incident side, and in the glass they are absorbed and created by emission, again and again.

The same happens by reflection on the surface of a mirror, but in this case the light is emitted to the same side from which it hits the surface – or in a direction under a certain angle relative to the incident light. So, the physics of the transparent body is not difficult to understand.

More difficult to understand is the fact that there also exist non-transparent bodies. In such bodies the molecules do not succeed in emitting the excitation energy totally as an electromagnetic wave; part of the energy is lost by collisions with neighbour molecules, leading to the production of heat. Whenever a non-transparent body absorbs light, it warms up.

Also, the colors of a body are understandable in this picture. Most of the molecules are picky, particularly in absorption of light. What energy is not lost by collisions with neighbouring molecules is emitted or reflected. When the light is emitted in the surface layer into the direction the light is coming from, we call it reflection. The white light from the sun consists of all the colors we see in the rainbow: red, orange, yellow, green, blue, violet, and the whole variety of the nuances between these colors. A body absorbing (and not emitting again) mainly the blue in the white incident light appears yellow because the rest of the light – white light less blue light – is yellow. Thus, the lemon is dark in blue light; it cannot emit yellow light because there is no yellow in blue light. Therefore, it stays dark. Some other substances are able to absorb blue light and then emit yellow light. Such colors are the so-called fluorescence colors. Fluorescent substances also shine in blue light. The lemon is not able to do that.

8th day, about laws of nature.

Whenever we hear about a law, we may ask, "Why is that so?" Newton's law of gravitation is an inverse square law. It says that the gravitational force between two objects with mass decreases with distance proportional to the distance squared.

The same is true for the force between two electric charges although these are two completely different kinds of forces. The inverse square law is related to the fact that our space is three-dimensional.

There exist several concealed interrelationships in nature. It seems that we still are missing something like a higher lookout from where the laws of nature are more self-evident and easier to understand.

Maybe one should not speak about "laws". Neither the electrons nor the planets obey some law by their movements on orbits. The word "laws" implies too much the association that there is somebody decreeing rules according to which these objects, whether electrons or planets, have to behave. On the contrary, when we examine the features of the electron we see that the electromagnetic field is not a force to which the electron has to pay obeisance; it is rather a feature of the electric charge. In the same way gravitation is rather a feature of the mass or of the space surrounding the mass.

We create ideas of the features and attributes of the electron, but they are not necessarily compatible with each other. For each feature we create a model of the electron, a model that explains or describes this special feature. But all these models do not fit together; they don't result in a complete and consistent picture of the electron.

Therefore, the electron is only a name for many phenomena we are able to observe from a special perspective.

A higher lookout – as mentioned previously – is, for example, the theory of relativity. The principle of relativity was already known in the mechanics of Newton. The idea that only *relative* velocities existed was there. There was no experiment with which we could say whether a system was resting or instead was moving with constant velocity.

We know that from our experience with the train. We are not able to tell whether the train is resting or moving at a constant speed by throwing balls or by working with pendulums or with any other mechanical devices. Only by looking out of the window can we see that the train is moving. And even then we don't know if the train is moving on the resting earth, or if the whole earth is moving under the resting train in the opposite direction. All this was summarized by the statement, "Only relative velocities exist." No mechanical device exists which can measure an absolute velocity.

Then came the idea with the ether and with the light velocity as we have seen in the 2nd day. Michelson wanted to measure the absolute velocity of the earth with an optical device. When this failed we achieved the higher lookout of relativity theory by the statement: there is also no optical device with which an absolute velocity can be measured. There doesn't exist any device at all of that kind. It is senseless to speak about an absolute velocity.

On the 1st day we saw the difference between the field-free mass and the electromagnetic mass of the electron. The field-free mass is the mass of the electron without electric charge – if it makes sense to speak about that, like speaking of color without light. If there really are two masses, we cannot say anything about their individual values; we only know the sum.

Then we saw the difference between the resting mass and the relativistic mass; that was on the 4th day. The photon has no rest mass; otherwise it couldn't travel with light velocity. At light velocity it has a mass higher than zero but not infinite.

One of the big questions now is, "What mass is that – which kind of mass has the photon; is it field-free mass or electromagnetic mass?" After all, photons are created by the annihilation of the electromagnetic mass of the positronium. Or is it something else, a third kind? We know that the mass of the photon is weighty, i.e., it is influenced by gravity. The deflection of light from distant stars has been measured (first in 1919) when it is coming close to the sun and found to be attracted by the gravity field of the sun. At a total solar eclipse – when the moon hides the sun – the stars in the close neighbourhood of the sun can be seen. By comparing a photo of the stars around the hidden sun with a photo of the same stars taken at night – when the sun is on the other side of the sky - one can see that stars seemed to have moved away from the sun during the eclipse.

When you draw tracks of the starlight close to the sun you will realize that this mowing away was caused by the attraction of the sun, by the deflection of the light towards the sun. The light path was twisted in a way that made it appear to us as if the stars had moved away from the sun.

When, on the 1st day, we talked about mass we were tacitly using two different properties of mass. First, we saw the property that masses attract each other. We call this *gravitational* mass. The other property is the resistance each mass has against a change of its state of movement, acceleration, or a slowing down. Whenever a body with mass is at rest or is moving uniformly, we have to apply a force in order to change its condition, i.e., in order to move it, accelerate it, or slow it down. Force is required because the body has inertia, or the body has *inertial* mass.

The fact that the values of the two masses are equal is not self-evident. The equality was first shown to a high accuracy in Budapest by *Lorand Eötvös (1848 – 1919)* in 1909. A body could, in principle, have a gravitational mass reacting to the attraction of another gravitational mass and, in addition to that, a different value of inertial mass for resisting a velocity change. We saw that the electron has both mass and charge. The mass resists any velocity change, and the charge reacts to an electric field. These are two different attributes of an electron, whereas the two mass attributes, the gravitational mass and the inertial mass, are identical. There exist bodies with mass but without charge (e.g., the neutron) and also bodies with much charge but little mass (the electron). Bodies without any mass at all having only charge are not known, curiously, but the ratio between mass and charge varies very much. (The proton, with almost 2,000 times the mass of the electron, has the same amount of charge.)

That is different with the gravitational mass and the inertial mass. The ratio is always the same. The fact that the gravitational mass is equal to the inertial mass is not understandable in the framework of classical physics. In relativity theory this is expressed by the statement, that it is the same mass, which has both these properties. That is the so-called *equivalence principle* of relativity theory.

That doesn't seem very interesting, but it implies some very interesting things. This equivalence between the gravitational and inertial masses means, for example, that we cannot distinguish whether we are placed in the gravity field of a big mass or we are sitting in a space shuttle that is accelerated upward. Both have the effect of pushing us to the floor. This is also a kind of relativity. There is no device existing with which we could decide whether we feel a gravity field or an acceleration – if we do not look out of the window; but that is not fair. By the way, even then we could not tell whether the whole environment we were seeing was placed in the gravity field or if it was accelerated together with our shuttle.

This equivalence principle is one of the foundation pillars of general relativity theory. Relativity theory is divided into two parts. The first part, *special relativity* theory, deals with uniform velocities.

The second part, *general relativity* theory, deals with accelerations and forces.

The conception of force has changed with time – as has the picture of the electron.

Newton's conception of force – the force of gravity – was something acting *instantaneously* at long distances, e. g., between earth and moon. Since this is not in agreement with relativity theory, a new conception was developed, a field around every mass transmitting the force with light velocity. Not until we had quantum theory was the conception of the force as an *exchange of particles* developed.

Force being an exchange of particles sounds curious but it works, and for both gravity and electrostatics.

Imagine two neighbouring rowboats end to end, in each boat a person. The two skippers now start to throw big heavy stones to each other. One throws and the other catches. Next, the other throws and the first catches. As that play continues the number of stones is constant – in fact you only need one stone, which is thrown back and forth. The two boats are exchanging stones. And what is happening? The two boats are starting to drift apart. Each time a stone is thrown from one boat to the other the stone also transports momentum from one boat to the other boat – a momentum pushing the boats apart. That means the exchange of bricks by throwing from one boat to the other has the effect of a repulsive force.

If, on the other hand, the two skippers don't throw the stones but instead each skipper, with a very long arm, removes a stone from the other boat and pulls it into his own boat, what will then happen? Well, the two boats will drift together. Each time a stone is pulled into a boat that boat is pulled toward the other boat. – In this case the exchange of a stone also transports momentum, but it has the effect of an attraction rather than a repulsion.

The exchange of particles with mass indeed has the effect of a force, repulsive or attractive, depending on how the exchange is done.

The electrostatic force between electron and positron – between all particles with electric charge – is mediated by the exchange of photons, light quanta. This holds for the repulsive force between similar charges and for the attractive force between opposite charges. The photon can manage this although it has no rest mass because it has a relativistic mass. We again see how the three particles, electrons, positrons, and photons act jointly.

Also, gravitation is thought of as an exchange of particles, gravitons. Until now gravitons exist only in theory. Nobody has found them, but they are intensively searched for. Like the photons, they should have no rest mass. But since they travel with light velocity they should have a relativistic mass with which they can transmit a force – in this case an attractive force between two masses.

With some of the natural laws, maybe even with most of them, one has the feeling that they may have been formed arbitrarily. Take for example the famous law of energy conservation. A term or expression is defined from the product of mass and the square of velocity or from the product of a distance and a force. The physicist claims that these products are constant. We are not going to discuss the exact formulation. Why do we take just these products? It is a similar thing with momentum, as we saw on the 2nd day with the billiard balls. For momentum, the product of mass and velocity, there exists a law of conservation. The same is true for angular momentum. That may be confusing for the layman. The physicist has gotten used to it. But he is not entirely content with it. He still is seeking the higher outlook, the more general principles from which these laws can be deduced.

General principles are often the more insightful the more general they are. Such insightful principles are, for example, *symmetries* in nature. If we succeed in describing nature from such insightful, self-evident, principles then we have gained a good deal of understanding.

And indeed, there is a deep connection between the laws of conservation and symmetries.

Symmetries play a fundamental role in nature. One symmetry seeming self-evident to us is the symmetry with respect to time or invariance against time shift, also called homogeneity of time. That simply means that nature today behaves exactly as it did yesterday. The architects building the pyramids in Egypt had to fight against the same gravity force as architects have to fight today. It turns out that the conservation law of energy follows from this invariance against time shift.

Another self-evident symmetry is the homogeneity of space, the invariance against spatial shift. That simply means that nature is the same in the house of our neighbour as in our own house. And it turns out that the conservation law of momentum follows from this invariance against a movement from one place to another – of course with the same conditions.

Both symmetries, the homogeneity of time and of space, seem self-evident. Otherwise, could you imagine the mess you would have if the pc runs perfectly in the store and not at home (sometimes this indeed happens, and nobody knows why) or if the efficiency would depend on the day of the week.

From a third kind of symmetry, the isotropy of space, that is, the invariance against a change of direction, follows the conservation law of angular momentum. If no external fields (gravitation for example) exist then no direction is favored, i.e., in the space far from any planets or stars there is no favored direction. In our galaxy or in our solar system or on our earth we have locally favored directions. On our earth, for example, the up-down direction and the north-south direction due to the rotation of our globe are favored directions. This can be taken into account in calculations with angular momentum, e.g., in weather forecast calculations. These calculations, which use the conservation law of angular momentum, are also based on the fact that there is no favored direction beyond locally favored directions.

It is a great improvement of understanding that we can come to a description of nature from such general principles.

A strange thing is that we find these connections in pairs – energy and time, momentum and location, angular momentum and direction – and these pairs are just the quantities occurring in the Heisenberg uncertainty relation. We find these relations by combining space and time into a relativistic unit and together with that by combining the quantum theoretic quantities describing momentum and energy. This gives a formalism where all fit nicely together.

Finally, we come back to the scattering experiments with electrons that we saw on the 1st day.

There we sent electrons through two slits close together and saw on the screen behind the slits every single electron coming like a bullet and causing a flash on the screen. And we saw how, from all these single flashes, an interference pattern emerged. The interference pattern showed that the electrons passed through the two slits like waves. The single flashes showed that the electrons behaved like particles.

But if they are particles they cannot create interference patterns. And moreover, if they are particles they can pass through only one of the two slits.

Therefore, we shall try to trick the electrons into showing us whether they are particles or waves. We shall do that by observing each electron on its path through a slit.

We install a tiny lamp directly behind the slits so we can see the electron when it is passing through one of the slits. We hope to see through which slit the electron passed. Then the electron hits the screen and we can study the interference pattern – hopefully.

And it works terrifically. We find out through which slit every single electron passes and we see the electrons hitting the screen.

But something strange is happening. On the screen no interference pattern is seen. We only see an accumulation of flashes in the center of the screen and some few flashes out towards the edges of the screen – exactly as expected with particles. The interference has disappeared. The electrons behaved like particles.

But we have learned what the reason is: the lamp is disturbing the electrons. The lamp emits photons, and we only see an electron if it collides with a photon and if this photon is detected, i.e., hits our eye. By colliding with a photon the track of the electron is changed and the interference pattern is destroyed.

Therefore, we take a weaker lamp and install this behind the slits. We have two possibilities: we can take a lamp that emits fewer photons, or we can take a lamp that emits photons of lower energy.

We first try it with a lamp emitting fewer photons. And indeed, the interference pattern appears again. But it is not so clear. It is mixed up with the picture we had before, showing the electrons as particles. By looking closer at the story we discover that we don't see all the electrons behind the slit. Because the lamp emits too few photons we don't illuminate every electron. We realize this because we see flashes on the screen when no electron is seen behind the slits.

We now have a mixture of electrons we have seen and can tell which of the slits they passed, and of undisturbed electrons we haven't seen and cannot tell which slit they passed.

The first set of disturbed electrons creates the picture on the screen identifying themselves as particles. The second set of undisturbed electrons creates the interference pattern, which identifies themselves as waves. And we see the mixture of both these pictures on the screen.

So we didn't get anywhere. Whenever we register the electrons as particles they show the picture on the screen that particles show. Only the undisturbed electrons, where we dispense with the information about which slit they passed through, create the interference pattern.

Therefore, we shall try the whole process again, this time with a lamp that still emits many photons but each with lower energy, that means light with longer wavelength, red or even infra-red light. Then the photon energy is so low that it won't disturb the electron, and these electrons should create the interference pattern.

And indeed, it works. The interference pattern appears on the screen, bright and beautiful, without the overlay of the particle picture. And we also see every single electron behind the slits by a red flash. However, we are not able to tell through which of the slits the electron goes. The red flash is too blurred. We only see a diffuse red light behind the slits not assignable to one of the two slits. The wavelength of the light is too long for making a sharp picture of the electron.

With this arrangement we gained the knowledge that the interference pattern appeared only by renouncing the information about the track of the electron. Whenever we register the electrons as particles they behave like particles.

Now we shall try a last possibility to trick the electrons, to see if the interference is caused somehow by interaction between the electrons. We send so few electrons through the slits that we can be sure they don't disturb each other. We can do that by observing the flashes on the screen. Only after we see the flash on the screen from one electron do we send the next electron through the slits. Of course we have to wait a long time before we can see if all the flashes create a pattern on the screen. But we are sure that the electrons don't interact. They pass through one of the two slits but they pass one at a time. We don't know which of the slits the electrons pass through, but we can be sure each electron has to pass one or the other slit – if it is a particle.

And after some time we recognize the interference pattern on the screen. It is the pattern of waves passing through two slits, as if they would interact like waves although they pass the slits at different times.

We can try whatever we want. If we treat the electrons as particles, i.e., if we obtain information about their tracks, then they behave like particles. If we treat them as waves, i.e., if we renounce the information about their tracks, then they behave like waves. But we cannot get both attributes – the information of the track, the particle attribute, and the interference pattern, the wave attribute – at the same time.

That is completely illogical for our feeling. Perhaps that is caused by our conception that an electron is something which keeps its identity. If an electron arrives at the slits, then according to our understanding it has to pass through one of them. But in fact, what happens is: We send an electron in the direction of the two slits; a little later we register an electron at the screen. Of course, we think it is the same electron – and that may be the logical mistake we make, the reason for all these logical difficulties. In reality we don't know what happens between the moment we send the electron and when it strikes the screen. It seems as if the electron would emerge at the moment it is observed. We already saw (on the 5th day) how questionable the identity of the electron is.

The quantum mechanical description only contains what we just have stated. We send an electron to the slits and calculate the probability that an electron will appear on the screen at a certain point. That yields the interference pattern. What happens between slits and screen is unknown, maybe in principle unknown.

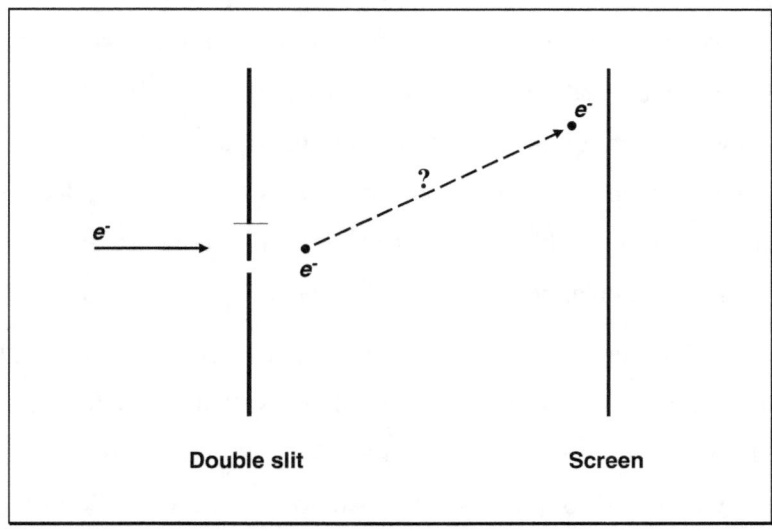

Fig.8.1. The identity of the electron.
Whenever we register an electron behind the two slits and a little later an electron at the screen we normally suppose it is the same electron.
In fact, in quantum mechanics we calculate the probability that an electron will appear on the screen at a certain point with the constraint that an electron appears behind one of the slits.
What happens in between is unknown. We don't know if it makes sense at all to speak about the same electron at the slits and at the screen.

If we detect something between these two events, for example gain information about which slit the electron passes, we have a totally different situation. Then we calculate the probability that an electron will appear on the screen at a certain point with the constraint that an electron appears behind one of the slits (Fig.8.1). We have to be content with that. The electron itself does not know more – or the phenomenon, which appears to us as electron, doesn't know more.

Now that the electron has disappeared we also finish our walk. The electron occurred at the beginning of our walk as a tangible small ball. Then it showed some strange features, almost bad habits according to classical physics. We had to tear down buildings of conception and reconstruct them. The electron became more and more diffuse. It did not know whether it should appear as a wave or as a particle. It got relatives, the positron and the photon. They also behaved very weirdly.

Finally, the electron lost its identity; only a phenomenon remained of which attributes first arose with observation– as if we would interpret those attributes as the phenomenon – an unreal ghost on the border between reality and non-existence. Nevertheless, this ghostly electron rules the entire technology of our everyday life. Light, communication, transport, and several comforts we owe to the electron – although we don't know what it is.

All that belongs to the history of the electron. On our walk we have casually touched most of modern physics – the principles of mechanics, including relativity theory, the nature of light, the principles of electrodynamics, the physics of the atom and quantum mechanics.

Of course, physics is a much broader field, but we have learned the structure on our trip – so that we can go deeper into the subject.

Why is nature that odd and according to our feelings so illogical? Maybe it is that the world of classical physics is not viable. We have seen all the difficulties coming from the classical view, from the classical concept of the electron.

We will finish our walk here. But we are standing in the midst of the landscape. The history of the electron is a continuing story – it is not yet finished. That we have to leave to later generations of scientists. Surely there will come more fascinating chapters and possibly it will remain incomplete forever.

References.

Some of the references are historically interesting (L. de Broglie, P. Jordan) and some of them are very helpful for physic students (P. Dirac, R. Feynman) and others are thought for the interesting layman. It is also profitable to meet the great physicists "personally" (W. Heisenberg, U. Röseberg, E. Segrè):

1. L. de Broglie, Matter and Light,
   Paris: Albin Michel, 1937
2. P. Davies, God and the New Physics
   J.M. Dent (1983) U.K.
   Simon & Schuster (1983) U.S.A.
3. P. Davies, Superforce
   Heinemann (1984) U.K.
   Simon & Schuster (1983) U.S.A.
4. P. Dirac, The Principles of Quantum Mechanics,
   Oxford University Press 1958
5. R. Feynman, The Character of Physical Law, The 1964 Messenger Lectures, MIT Press, 1967, ISBN 0-262-56003-8
6. R. Feynman, QED, The Strange Theory of Light and Matter, Princeton University Press 1985
7. R. Feynman, R. Leighton, M. Sands, The Feynman Lectures on Physics, Caltech 1961 - 1963
8. J. Gribbin, In Search of Schrödinger's Cat, quantum physics and reality, Black Swan Books, London 1996.

9. A.Guth, The Inflationary Universe, Helix Books, Addison-Wesley Publ. Comp., Massachusetts 1997

10. W. Heisenberg, Der Teil und das Ganze (The Part and The Whole), Piper Verlag, München 1969

11. N. Herbert, Quantenrealität (Quantum Reality: Beyond the New Physics), Birkhäuser Verlag, Basel 1987

12. P. Jordan, Anschauliche Quantentheorie, Springer Verlag 1936

13. U. Röseberg, Niels Bohr, Wissenschaftliche Verlagsgesellschaft, Stuttgart 1985

14. E. Segrè, Die großen Physiker und ihre Entdeckungen, Piper Verlag, München 1984

15. F. Selleri, Die Debatte um die Quantentheorie (Le grand débat de la théorie quantique), Vieweg, Braunschweig 1983

www.ingramcontent.com/pod-product-compliance
Lightning Source LLC
Chambersburg PA
CBHW051709170526
45167CB00002B/592